AC DC

e E V electric potential Volts (V)

i I "   current   Amps (A)

    R "   resistance   Ohms (Ω)

voltage measured parallel	$V_{rms} = \frac{V_{peak}}{\sqrt{2}}$
Current " series	
R " isolation	$I_{rms} = \frac{I_{peak}}{\sqrt{2}}$

Power   DC $P = V \cdot I = \frac{V^2}{R} = R I^2$

AC $P_{AC} = V_{rms} \cdot i_{rms} = \frac{V_{rms}^2}{R} = R i_{rms}^2$

$P_{AC} = V_{rms} \cdot i_{rms} = \frac{V_{peak}}{\sqrt{2}} \frac{i_{peak}}{\sqrt{2}} = \frac{P_{peak}}{2}$

$P_{AC} = \frac{V_{peak}^2}{2R} = \frac{R i_{peak}^2}{2}$

Solenoid

$F = \frac{1}{2}(NI)^2 \frac{\mu \cdot A}{x^2}$ (pg 51)

pilot spool

$F_x = F_i + B_x$ (pg 69)

$B = \frac{F_f - F_i}{S}$, force gradient $\left(\frac{lbf}{in}, \frac{N}{cm}\right)$

$t = \sqrt{\frac{M}{B}} \ln\left(\frac{x + r + \sqrt{2rx + x^2}}{}\right)$ (pg 71)

$V = \frac{Q}{A}$ pg (72)

$Q = CA \sqrt{\frac{2 \Delta p}{\rho}}$ (p 72)

$i = \frac{V}{R_1} + \frac{V}{R_2} = \frac{V}{R_{eq}}$

$\frac{1}{R_{eq}} = \frac{1}{R_1} + \frac{1}{R_2}$

$1 \, gal = 231 \, in^3$

$1 \, lbm = \frac{1}{32.17} \frac{lbf \cdot s^2}{ft} = \frac{1}{386.1} \frac{lbf \cdot s^2}{in}$

$1 \, kg = \frac{N \cdot s^2}{m} = \frac{1}{100} \frac{N \cdot s^2}{cm}$

$WP = \frac{BP}{SF}$ working pressure = Burst pressure / factory of Safty.

pressure loss $\Delta p = K \frac{1}{2} \rho V^2$ (K is called the loss coefficient)

Resistor Color Code

Color	Value		Tolerance %		First band: First sig fig
Black	0	Gold	5		Second band: 2nd "
Brown	1	Silver	10		Third band: Multiplier
Red	2	No band	20		4th band: Tolerance
Orange	3				gold .1 → multiplier
Yellow	4				Silver .01 → multiplier
Green	5				
Blue	6				
Violet	7				
Grey	8				
white	9				

$$PSIG = (PH_{ft})(SG)(.433)$$

$$in\text{-}Hg = (NH_{in})(SG)(.0736)$$

$$A = \frac{\pi d^2}{4}$$

$$F = P \cdot A$$

$$PSIA = 14.7 + (NH_v \cdot 0.4912) \quad \text{(in Hg vacc)}$$

$$V_a = (Q)(231)\left(\frac{\#}{60}\right) \quad accum.$$

$$(P_1)(V_1) = (P_2)(V_2)$$

$$F_{regen} = P \cdot A_r$$

Decompression of oil

$$V = A \times S$$

$$DV = (\%C)(V)$$

Electrohydraulic Control Systems

F. Don Norvelle

Oklahoma State University

Prentice Hall
Upper Saddle River, New Jersey   Columbus, Ohio

Library of Congress Cataloging-in-Publication Data

Norvelle, F. Don.
  Electrohydraulic control systems/F.D. Norvelle.
    p. cm.
Includes bibliographical references and index.
ISBN 0-13-716359-2
1. Hydraulic control. I. Title.

TJ843 .N57 2000   •
629.8'042--dc21
                            99-053027

Publisher: Charles E. Stewart, Jr.
Production Editor: Alexandrina Benedicto Wolf
Production Coordinator: Karen Fortgong, *bookworks*
Cover Design Coordinator: Karrie Converse-Jones
Cover Designer: Jeff Vanik
Cover photo: © Michael Simpson, FPG International
Production Manager: Matthew Ottenweller
Marketing Manager: Ben Leonard

This book was set in Times Roman by York Graphics and was printed and bound by R. R. Donnelley
& Sons Company. The cover was printed by Phoenix Color Corp.

Prentice-Hall International (UK) Limited, *London*
Prentice-Hall of Australia Pty. Limited, *Sydney*
Prentice-Hall of Canada, Inc., *Toronto*
Prentice-Hall Hispanoamericana, S. A., *Mexico*
Prentice-Hall of India Private Limited, *New Delhi*
Prentice-Hall of Japan, Inc., *Tokyo*
Prentice-Hall (Singapore) Pte. Ltd., *Singapore*
Editora Prentice-Hall do Brasil, Ltda., *Rio de Janeiro*

CONTENTS

PREFACE

This textbook is aimed specifically at students enrolled in two- and four-year college-level programs in engineering technology, especially in those programs where motion control is taught. Its primary objective is to provide the student with a very broad introduction to the concepts of electrohydraulic control systems specifically and motion control in general, and to do so in a one-semester course. It also attempts to bridge the gap between the traditional electronics and mechanical curricula.

The second objective is really not new. There are several texts that attempt to close the gap in specific areas. Servomechanisms and proportional control valves are good examples of electrohydraulic systems for which there are several bidiscipline references. These texts, however, are highly theoretical and are intended more as references for engineers than as textbooks for engineering technologists. There are also numerous texts on control theory, but these tend to deal almost entirely with electronics.

In this text I have attempted to present both the mechanical and electronic aspects of the various types of valves and illustrate methods of controlling those valves. It is important to note that although the title of the text implies that it is specifically concerned with hydraulic systems, most of the concepts presented in the discussion of controls (ladder diagrams, programmable logic controllers, sensors, and operational amplifiers) are applicable to all motion systems. Only the power sections of the systems are different.

This text assumes a basic knowledge of fluid power, so Chapter 2 provides only a brief review of that subject. A full treatment of hydraulics is well beyond the scope of this text, and is, in fact, the subject of numerous textbooks and college courses.

Chapter 3, likewise, assumes a basic knowledge of electrical circuitry, so the first part of the chapter offers only a brief review of the subject. However, many of the electronics courses taught in current curricula are very heavy on electronics and light on electrical circuitry. For this reason the concepts of common electrical switches and relays are introduced in this chapter. The chapter then continues with a fairly detailed discussion of ladder diagrams. This discussion lays the groundwork for solenoid valve control circuits in Chapter 4 and PLC programming in Chapter 9.

Chapters 5 and 6 deal with proportional control valves and servovalves, respectively. In both chapters the mechanical aspects of the valves, the valve actuation mechanisms, and the control circuit boards are discussed. Valve performance is also discussed.

The calculations for determining the natural frequencies of hydraulic motor and cylinder systems are demonstrated in Chapter 7. This chapter also includes the concepts of gain, transfer functions, and feedback and control using a "running" example.

Chapter 8 expands on the basic switching elements that were introduced in Chapter 3. Sensor properties such as accuracy and precision are addressed, along with analog to digital conversion. There is extensive discussion of discrete sensors, including proximity, photoelectric, Hall effect, and reed switches. Continuous sensors for temperature, flow rate, pressure, linear and rotary position, and velocity and acceleration are discussed as well.

Programmable logic controllers are the subject of Chapter 9, which begins with an introduction to the basic concepts of PLCs, architecture, and programming languages. A section on ladder logic programming follows, which facilitates the transition from the relay logic diagrams of Chapters 3 and 4 to the ladder logic required for programming PLCs. The chapter ends with some programming examples.

The final chapter (Chapter 10) is a brief introduction to robotics. It discusses robot geometries, end effectors, and control and programming.

Obviously, this is very much an introductory text. The subject of virtually every chapter has been treated extensively in numerous books. The purpose here is simply to introduce the student to the concepts in considerable breadth but limited depth. To my knowledge, this is the first textbook to approach electrohydraulic control in this manner.

Some chapters include more depth than might be desired for a specific curriculum. The structure of the text caters to those cases. Each chapter begins with basic descriptions and proceeds to more analytical material. It is a simple matter, then, to use only specific parts of a chapter if the depth is not needed for the course.

The material is generally presented in a fashion that facilitates associated laboratory exercises where lab facilities are available. For example, the solenoid valve circuits presented in Chapter 4 lend themselves nicely to basic hydraulic training stands. The PLC programming examples can be used directly as laboratory exercises. (Where PLCs other than the OMRON are used, it would make sense to translate these examples into the appropriate programming language.)

A unique feature that will help demonstrate the relationship between the hydraulic circuit and the control circuit is the inclusion of the Automation Studio CD-ROM with the instructor's manual. This software allows the student to draw both the hydraulic circuit and the electrical ladder diagram and then simulate the system operation. When the START button is "pushed" on the screen, the appropriate electrical components are energized (change color), the valves move, and the cylinder or hydraulic motor operates. It is an easy transition, then, to take the proven circuit from the "drawing board" to the training bench.

This text will not make the student a controls expert, but it will provide a solid basis on which to build. Industry badly needs people with this type of academic background.

ACKNOWLEDGMENTS

So many people have contributed to the preparation of this book that I am almost afraid to start listing them for fear of overlooking someone. Special thanks, though, must be given to several people: Frank Garner of the Vickers Training Center; Randy Nobles of Womack Machine Supply in Tulsa, Oklahoma; Tom Nelson and Rob Bartling of Racine Federated; Glenn Haueter of Automated Dynamics in Tulsa; Larry Schrader of the Parker Fluid Power Training Center; and Jim Green of Control Dynamics in Oklahoma City.

These people, with the permission of their respective companies, have provided literature, hardware, and even opportunities for training that have allowed me to gain the knowledge, understanding, and experience necessary to write this text. A special acknowledgment goes to Stewart Atkin of FAMIC Technologies 2000 of St. Laurent, Quebec, Canada, for providing the Automation Studio Demo CD-ROMs.

My heartfelt thanks go to my wife, Evelyn, who got so tired of watching me struggle with the keyboard that she pushed me aside and typed (and retyped) the manuscript. I also appreciate the efforts of my daughters, Babs and Cery, who proofread the manuscript and very seldom laughed (out loud, anyway) at my mistakes. I also thank my electrohydraulics students at Oklahoma State University for patiently struggling through the manuscript. Their suggestions were very useful.

I also thank the following who reviewed the manuscript and made so many excellent suggestions: Ciso Macia, Arizona State University; David Pacey, State University of Kansas; Bill Reeves, Ohio University; Wajiha Shireen, University of Houston; and Gang Tao, University of Virginia. I took those suggestions to heart and incorporated most of them.

Finally, my sincere thanks to the staff of Prentice Hall Publishing Company. They were very patient with me and provided invaluable help and encouragement.

F. Don Norvelle

CHAPTER 1

An Introduction to the Real World

1.1 INTRODUCTION

Fluid power is involved in virtually every phase of industry, including manufacturing, transportation, and construction. Hydraulic systems are required in heavy-load applications, whereas pneumatic systems are generally employed in light-load, short-stroke, high-speed applications. In many cases, simple manual control systems are totally adequate for the operation, even where a considerable amount of complexity is required. However, the "real world" of fluid power lies in the domain of electric and electronic command, control, and sensing. The marriage of electronics and hydraulics has produced a hybrid system that has both brains and brawn. The electronics can utilize digital devices such as limit switches, pressure switches, and low-level switches to provide nearly instantaneous reaction to limiting situations. The use of tachometer-generators, linear variable differential transformers, piezoelectric pressure sensors, turbine flow meters, accelerometers, and other continuously sensing devices with electronic outputs allows very precise control of both linear and rotational speed, position, acceleration, and deceleration as well as force. The hydraulic system concomitantly provides the same stiffness, accuracy, power, and reliability for which it has always been known.

With the advent of programmable logic controllers (PLCs), the scope of automating with electrohydraulics has become virtually unlimited. The concept of a fully automated manufacturing facility controlled from a single location is now a reality. With PLCs, not only is the manufacturing process control fully automated, it is also readily changed. Previously, many types of piece-part manufacturing (gears, for instance) were limited to large-batch quantities because of the changes that were required from one batch to the next. That is no longer the case. With all the processes stored in the PLC memory, if a single gear of one type is needed, the operator need only call up that program and load the gear blank.

Of course, PLC control is not limited to hydraulics. Every phase of the process—heat treating temperatures and times, surface finishes, painting, packaging—can be handled by a PLC network using a master computer with slave or satellite units for the individual processes.

1.2   APPLICATIONS

Even the simplest industrial systems use electrohydraulics to cause cylinders to stop at a certain position, reciprocate automatically, or operate in a certain sequence. The most common method for achieving these functions is to use solenoid valves activated by limit switches or some other type of digital device. Figure 1.1 shows a bank of solenoid valves that control the operation of cylinders in a manufacturing process.

The use of electrohydraulics in mobile and construction equipment enhances operator safety and reduces operator fatigue while providing improved controllability. The use of a joystick to control electrically actuated valves located on or near the output power actuators removes the high-pressure hydraulic lines from the operator's station and reduces the physical effort that was previously required to manipulate large valve handles. The rock drill in Figure 1.2 utilizes this type of system.

Figure 1.1 A bank of solenoid-operated valves.

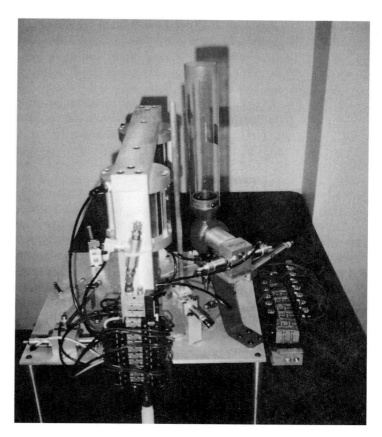

Figure 1.2 A hydraulically operated rock drill. (Courtesy of P&H Mining Equipment)

Modern high-performance aircraft commonly use "fly-by-wire" flight control systems. In these aircraft the pilot, rather than operating valves to port fluid to the control-surface actuators, sends electrical signals to a flight control computer. These signals tell the computer what the pilot wants the airplane to do. The computer then sends the necessary commands to the appropriate actuators to achieve the desired maneuver. Because most of today's fighter aircraft are aerodynamically unstable, they are virtually impossible to control by conventional methods; therefore, computer control is an absolute necessity. A typical aircraft hydraulic system is shown in Figure 1.3.

Other electrohydraulic applications include:

- Passenger-car leveling systems for high-speed trains.
- Railroad track sensing and alignment equipment for straightening bent rails.
- Roll-control systems for oceangoing ships.
- Ship propeller pitch and speed controls.

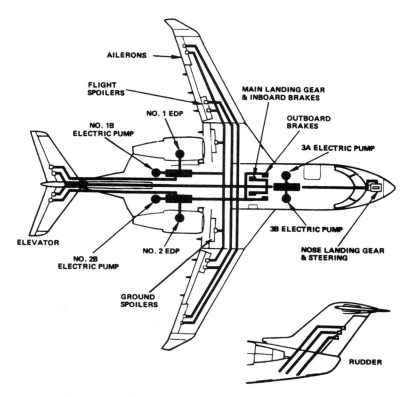

Figure 1.3   A typical aircraft hydraulic system powers the flight controls, landing gear, and nose wheel steering as well as other aircraft functions. (Courtesy of Neese, William A., *Aircraft Hydraulic Systems,* 1991, Krieger Publishing Company, Malabar, Florida)

- Steam valve control for electrical generators.
- Roll control for offshore drilling rigs.
- Automobile transmissions.
- Thrust vectoring nozzles for spacecraft.

The list is nearly endless. Virtually every machine that moves or is used for manufacturing has either hydraulics or pneumatics, and most have some sort of electrical or electronic control.

1.3   THE FUTURE

Fluid power is considered a "mature technology." That term implies that although there may be small, evolutionary advances, no revolutionary changes can be expected. Although this *may* be true, the future of electrohydraulics and electropneumatics is, nonetheless, exciting.

Certainly, evolutionary advances are being seen constantly. Researchers are continually working to improve standard solenoids to provide higher forces and faster operation. Proportional solenoid performance has already reached levels of control and frequency response that rival some servovalves, and the work is continuing. Command and feedback electronics for servovalves are advancing in step with all other areas of electronics, and the valve torque motors and the valves themselves are constantly being improved to keep pace with the electronics.

Improved transducers, digital electronics, and fiber-optic devices are making their way into the electrohydraulic world. "Smart" components, such as the servocylinder shown in Figure 1.4, servopumps, and servomotors are being developed. In these devices, the servovalve, the cylinder, the feedback unit, and the servo electronics are combined in a single package. The only external electrical connections are a power lead and a single command line from the master computer. The unit in Figure 1.4 is digitally addressed from the central command computer, which can control eight such units simultaneously.

All these advances, however, are evolutionary. So far, we have witnessed the marriage of separate electrical/electronic devices and fluid power devices—two separate sets of hardware combined to perform a specific function. The future will bring us the offspring of that marriage—a truly hybrid electrohydraulic device in which there will be no demarcation between the electronics and hydraulics. Rather than being simply a control valve bolted to a cylinder, a unit will be fully integrated, and the separate components will be virtually indistinguishable. *That* will be revolutionary and will open a new world for electrohydraulic applications.

I encourage you to be a part of the revolution. This book is a good start, but it is only the beginning. Other texts will be referenced in subsequent chapters that contain

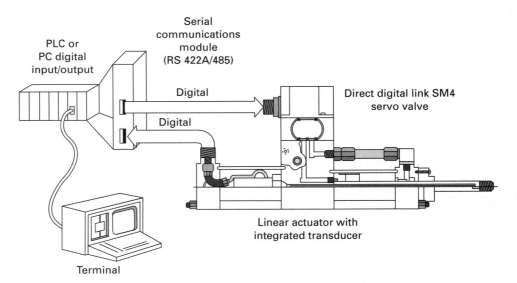

Figure 1.4 A hydraulic servocylinder with the servovalve and controller integrally mounted.

much more detailed information on the various devices. This text introduces you to the electrohydraulic components and their applications but makes little effort to develop theories or design concepts. I hope you will advance into those areas either through formal education or self-study.

Above all, remember that this book is based on *today's* technology and *today's* electrohydraulic concepts. What you will learn here represents the way things are done today, but there may be other, perhaps even better, ways. Look for those better ways. Be creative and imaginative. If this text, or any other text, interferes with your creativity, throw the book away! For now, however, use it as the foundation for your future growth in electrohydraulics.

CHAPTER 2

A Review of Basic Fluid Power

OBJECTIVES

When you have completed this chapter, you will be able to:

- Explain the concept of energy conversion throughout a fluid power system.
- Discuss the relationships of flow and pressure.
- Calculate cylinder and motor speeds based on flow rate.
- Calculate cylinder force and motor torque based on pressure.
- Calculate input, hydraulic, and output power.
- Recognize and identify the commonly used ISO (International Organization for Standardization) graphic symbols for fluid power components.
- Read hydraulic circuit diagrams drawn with ISO graphic symbols.
- Draw simple hydraulic circuit diagrams using ISO graphic symbols.

2.1   INTRODUCTION

Electrohydraulics is an efficient and useful marriage of two well-defined disciplines. To fully understand how this combination can be used to best advantage, it is important to have a basic understanding of each discipline. If you have come this far in your studies, you must have a basic understanding of the concepts of fluid power. In this chapter, therefore, we will take just a cursory look at fluid power. We will spend some time reviewing the basic concepts, then look at the most commonly used symbols and circuit types. The emphasis here is on hydraulics, although much of the material applies to pneumatics also.

2.2   FLUID POWER CONCEPTS

The objective of a fluid power system is to do useful work. This is accomplished in three fundamental steps. First, a mechanical energy input is converted into fluid energy by a hydraulic pump. Next, this fluid energy is transmitted through fluid conduits and any necessary control devices. Last, the fluid energy is reconverted into mechanical energy by an output device—usually a hydraulic cylinder or a hydraulic motor. This process is demonstrated in Figure 2.1, in which an electric motor drives a pump. The pump output provides the energy to drive a hydraulic motor, which drives a gear box and, subsequently, the load.

Notice the parameters along the top of this figure. The arrows indicate that there is a voltage input into the electric motor, which results in a rotating speed (rpm). The speed of the electric motor is the input to the pump and results in a flow (gpm) output from the pump. This flow, in turn, causes a speed (rpm) output from the motor. The motor speed determines the speed of the gearbox and the load. This diagram implies that the voltage input to the system controls the speed output of the system.

Now, look at the parameters along the bottom of the figure. The arrows here point from right to left, indicating that the load dictates these system parameters. Thus, the torque (lb·in.) required to turn the load dictates the torque needed to turn the gearbox. The gearbox input (which is the hydraulic motor output) determines the pressure (psi) required to turn the hydraulic motor. The torque required to turn the pump is the result of this pressure. The electric motor output torque governs the electric current (amps).

This diagram illustrates some fundamental concepts concerning flow and pressure. First, the purpose of a pump is to produce *flow*. Pump output is flow, *not* pressure. In fluid power systems we often require that the flow output remain constant (or nearly so) regardless of system pressure; therefore, we use positive displacement pumps almost exclusively. Often, these pumps are equipped with devices that allow us to change the volumetric displacement and, consequently, the output flow rate. These devices include manual adjustments, hydromechanical pressure compensators, and electronic controllers such as proportional solenoids or servo actuators.

If pumps produce flow, then where do we get pressure? That is the second fundamental concept: pressure is the result of resistance to flow. Resistance to flow comes from two sources—the load and the liquid. The resistance of the load is obvious. The resistance of the liquid results from the viscosity of the liquid, from friction as the liquid moves through the system conduits, fittings, and valves, and from changes in the direction of flow (inertial effects).

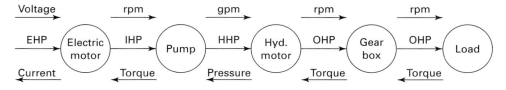

Figure 2.1 Energy flow in fluid power circuits.

The third fundamental concept is that the maximum pressure in any branch of a circuit is the result of the minimum resistance to flow in that branch. This concept will be demonstrated later in this chapter.

2.2.1 Flow–Speed Relationship

Yet another fundamental fluid power concept concerns the flow–speed relationship. Simply stated, the speed of operation of any hydraulic output device (whether a cylinder or a hydraulic motor) depends solely on the rate at which fluid is pumped into the device. It has nothing to do with pressure (assuming, of course, that there is sufficient system pressure capability to move the load at all). This is not as clear cut in pneumatics as in hydraulics, but the basic concept is true for both.

The velocity of a hydraulic cylinder is calculated from Equation 2.1:

$$v = \frac{Q}{A} \tag{2.1}$$

where v = cylinder velocity
         Q = flow rate
         A = effective area (A_P for the piston end, $A_P - A_R$ for the rod end)

Example 2.1: A double-acting, single-ended hydraulic cylinder has a 2-in. (5.08-cm) bore (or diameter) and a 1-in. (2.54-cm) rod. The system flow rate is 3 gpm (11.36 lpm). Calculate the extension and retraction velocities of the cylinder.

Solution: The extension velocity depends on the piston area, which is

$$A_P = \frac{\pi D_P^2}{4} = \frac{\pi (2 \text{ in.})^2}{4} = 3.14 \text{ in.}^2 \ (20.27 \text{ cm}^2)$$

Therefore, the extension velocity is

$$v = \frac{Q}{A_P} = \frac{(3 \text{ gal/min})(231 \text{ in.}^3/\text{gal})}{3.14 \text{ in.}^2} = 221 \text{ in./min} \ (561 \text{ cm/min})$$

The effective area on retraction is the piston area minus the rod area. The rod area is

$$A_R = \frac{\pi D_R^2}{4} = \frac{\pi (1 \text{ in.})^2}{4} = 0.78 \text{ in.}^2 \ (5.03 \text{ cm}^2)$$

Thus,

$$v = \frac{Q}{A_P - A_R} = \frac{(3 \text{ gal/min})(231 \text{ in.}^3/\text{gal})}{(3.14 - 0.78) \text{in.}^2} = 294 \text{ in./min} \ (747 \text{ cm/min})$$

From this example we see that for any given flow rate, the retraction speed is faster than the extension speed for any single-ended, double-acting cylinder. The ratio of the speeds is the same as the ratio of the effective areas on which the pressure acts. Therefore:

$$v_{\text{ret}} = \frac{A_P}{A_P - A_R}\, v_{\text{ext}} \tag{2.2}$$

Notice that no pressure term is involved in any of these velocity calculations.

The rotating speed of a hydraulic motor is also determined by the flow rate. Instead of being a function of the area, as in a cylinder, the speed is a function of the volumetric displacement of the motor. Equation 2.3 defines the relationship:

$$N = \frac{Q\eta_v}{V_M} \tag{2.3}$$

where N = rotating speed
Q = flow rate through the motor
V_M = volumetric displacement of the hydraulic motor
η_v = volumetric efficiency of the hydraulic motor

Example 2.2: Determine the rotating speed of a hydraulic motor with a displacement of 2.8 in.^{3}/rev (45.9 cm^3/rev) for a flow rate of 6 gpm (22.7 lpm). The volumetric efficiency of the motor is 0.92.

Solution: Using Equation 2.3, we have

$$N = \frac{Q\eta_v}{V_M} = \frac{(6\ \text{gal/min})(231\ \text{in.}^3/\text{gal})(0.92)}{2.8\ \text{in.}^3/\text{rev}} = 455\ \text{rpm}$$

2.2.2 Pressure–Force Relationship

The pressure that results from the application of a force depends on the area over which that force is exerted. Pressure is defined algebraically as:

$$p = \frac{F}{A} \tag{2.4}$$

where p = pressure
F = force
A = effective area

Example 2.3: A hydraulic cylinder with a 4-in. (10.16-cm) bore supports a 10,000-lb (44.5-kN) load as shown in Figure 2.2. What is the pressure on the gauge?

Solution: The area of interest is the area in contact with the supporting fluid—in this case, the entire area of the piston. Thus,

$$A = 12.56\ \text{in.}^2\ (81.07\ \text{cm}^2)$$

Figure 2.2 Example 2.3.

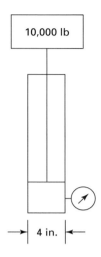

10,000 lb

4 in.

Therefore,

$$p = \frac{F}{A} = \frac{10{,}000 \text{ lb}}{12.56 \text{ in.}^2} = 796 \text{ psi } (5.5 \text{ MPa})$$

Example 2.4: A 4-in. (10.16-cm) hydraulic cylinder supports a suspended 10,000-lb (44.5-kN) load as shown in Figure 2.3. The rod diameter is 1 in. (2.54 cm). What is the pressure on the gauge?

Solution: Again, we use Equation 2.3 to find the pressure, but the area with which we are concerned is the annular area around the rod, that is, the area of the piston minus the area of the rod. Therefore,

$$A_N = A_P - A_R = 12.56 - 0.78 = 11.78 \text{ in.}^2 (76.01 \text{ cm}^2)$$

Figure 2.3 Example 2.4.

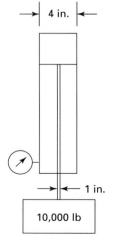

4 in.

1 in.

10,000 lb

So

$$p = \frac{F}{A_N} = \frac{10,000 \text{ lb}}{11.78 \text{ in.}^2} = 849 \text{ psi } (5.85 \text{ MPa})$$

Another way to look at this force–pressure relationship is to consider the force capability of a cylinder based on the maximum pressure available to it. This pressure availability is usually determined by the relief valve setting.

Example 2.5: Find the maximum extension force capability of a 3-in. (7.62-cm) cylinder if the system relief valve is set at 2000 psi (13.8 MPa).

Solution: From Equation 2.3, we get

$$F = p \times A = (2000 \text{ lb/in.}^2)(7.07 \text{ in.}^2) = 14,137 \text{ lb } (62.9 \text{ kN})$$

Note that this is the maximum force *capability*. The actual force depends on the load. Increasing the relief valve setting increases the maximum capability, but it does *not* change the actual pressure resulting from a specific load situation.

There is a pressure–force type relationship for hydraulic motors, also. Actually, the "force" is torque, and it is a function of the pressure differential across the motor rather than the actual upstream pressure. Torque is found from Equation 2.5:

$$T = \frac{V_M \Delta p \eta_m}{2\pi} \tag{2.5}$$

where T = torque
    V_M = volumetric displacement
    Δp = pressure differential across the motor
    η_m = mechanical efficiency

In most cases, the exhaust port of the motor returns directly to the reservoir. If the reservoir is unpressurized, we usually assume that the pressure at the motor outlet is near zero. Although this assumption may not be exactly correct, it is handy because it allows us to use the inlet pressure alone for calculating torque.

Example 2.6: A hydraulic motor with a 2.8-in.^{3}/rev (45.9 cm^3/rev) displacement has a mechanical efficiency of 0.94. The inlet pressure is 2000 psi (13.8 MPa). Find the torque produced.

Solution: We use Equation 2.5 to solve this problem. Since we are given no information about the pressure at the outlet port, we will assume that it is approximately zero. Thus,

$$T = \frac{V_M \Delta p \eta_m}{2\pi} = \frac{(2.8 \text{ in.}^3/\text{rev})(2000 \text{ lb/in.}^2)(0.94)}{(2\pi)} = 838 \text{ lb·in. } (94.7 \text{ N·m})$$

Keep in mind that we were able to calculate the torque because we knew the pressure. In reality, the requirement for the torque generates the pressure. Remember that pressure is a function of the resistance to flow. In this case, the resistance to flow is the load that creates the torque requirement. Remember also that there are many other resistances to flow, all of which result in higher system pressures. These resistances are primarily from flows through fluid conduits, fittings, and valves. Unfortunately, all these resistances represent inefficiencies in the circuit. The result of inefficiency is energy loss and heat generation. These are undesirable, so the good designer will select conduit sizes, layouts, and valve designs to minimize these problem areas.

2.2.3 Horsepower Calculations

Several different horsepowers are of interest in fluid power systems. These include:

   a. Pump input horsepower.
   b. Hydraulic horsepower (pump output).
   c. Hydraulic motor output horsepower.
   d. Hydraulic cylinder output horsepower.
   e. Horsepower losses through valves, fittings, lines, and the like.

The pump input horsepower is also the output horsepower of the prime mover (normally an electric motor or an engine of some kind) and is sometimes termed the *torque horsepower*. We will use the acronym IHP to denote this input. In the U.S. customary system of units, we find IHP from the equation

$$\text{IHP} = \frac{T \times N}{5252} \tag{2.6}$$

where T = torque (lb·ft)
        N = rotational speed (rpm)

The conversion factor is 5252 based on these specific units for torque and speed. If torque is given in pound-inches, the factor is 63,025.

In the SI system, torque is measured in newton-meters (N·m) rather than in pound-inches, and the conversion factor is 9550. Also, power is measured in kilowatts rather than horsepower, so Equation 2.6 becomes

$$P_{\text{in}} = \frac{T \times N}{9550} \tag{2.6a}$$

where P_{in} = input power (kW)
        T = torque (N·m)
        N = rotational speed (rpm)

The pump output horsepower is usually termed either hydraulic horsepower (HHP) or fluid horsepower (FHP). It is based on the measured system pressures and flow rates and is found from

$$\text{HHP} = \frac{p \times Q}{1714} \tag{2.7}$$

where p = system pressure (psi)
          Q = flow rate (gpm)

The conversion factor is 1714 based on those units.

In the SI system, pressure is measured in kilopascals and flow rate in liters per minute, so Equation 2.7 is written

$$P_{hyd} = \frac{p \times Q}{60,000} \qquad (2.7a)$$

where P_{hyd} = hydraulic power (kW)
          p = pressure (kPa)
          Q = flow rate (lpm)

The pump input and output horsepowers are related by the overall pump efficiency, which is simply the ratio of the output to the input. Thus,

$$\eta_o = \frac{HHP}{IHP} \qquad (2.8)$$

EXAMPLE 2.7: A hydraulic pump is operating at 1500 psi (10.3 MPa) and 10 gpm (37.85 lpm). The pump is driven by an electric motor at 1725 rpm. The electric motor produces 350 lb·in (3.95 kN·m) of torque. Find the overall efficiency of the pump.

Solution: The horsepower input to the pump is found from Equation 2.6:

$$IHP = \frac{T \times N}{63,025} = \frac{(350 \text{ lb·in.})(1725 \text{ rpm})}{63,025} = 9.58 \text{ HP } (7.15 \text{ kW})$$

We use Equation 2.6 to find the horsepower output of the pump:

$$HHP = \frac{p \times Q}{1714} = \frac{(1500 \text{ psi})(10 \text{ gpm})}{1714} = 8.75 \text{ HP } (6.53 \text{ kW})$$

Now, we find the overall efficiency from Equation 2.7:

$$\eta_o = \frac{HHP}{IHP} = \frac{8.75}{9.58} = 0.91 \text{ or } 91\%$$

The horsepower output from a hydraulic motor is also a torque–speed relationship:

$$OHP = \frac{T \times N}{63,025} \qquad (2.9)$$

or, in SI units,

$$P_{out} = \frac{T \times N}{9550} \qquad (2.9a)$$

The input horsepower is based on the pressure drop across the motor and the flow rate through the motor. We often express this value as hydraulic horsepower and calculate it from

$$\text{HHP} = \frac{\Delta p \times Q}{1714} \tag{2.10}$$

or, in SI units,

$$P_{\text{hyd}} = \frac{\Delta p \times Q}{60,000} \tag{2.10a}$$

where $\Delta p = p_{\text{inlet}} - p_{\text{outlet}}$

As we saw earlier, the outlet line returns the fluid to the system reservoir at a relatively low pressure. For this reason, we often neglect the outlet pressure and simply use the inlet pressure in the horsepower calculation. Note that although we call this quantity hydraulic horsepower, it is usually not the same as the pump output horsepower.

The overall efficiency of the motor defines the ratio of the output and input horsepowers:

$$\eta_{\text{o}} = \frac{\text{OHP}}{\text{HHP}} \tag{2.11}$$

Notice the similarity between this equation and Equation 2.8. Remember that in both equations we are relating output horsepower to input horsepower.

EXAMPLE 2.8: A hydraulic motor produces 150 lb·in. (16.95 N·m) of torque at 1600 rpm. Find the output horsepower of the motor.

Solution: Using Equation 2.9, we find

$$\text{OHP} = \frac{T \times N}{63,025} = \frac{150 \times 1600}{63,025} = 3.81 \text{ HP (2.84 kW)}$$

EXAMPLE 2.9: The motor of Example 2.8 has an input flow rate of 8 gpm (30.28 lpm), and there is a pressure drop across the motor of 875 psi (6.03 MPa). Find the overall efficiency of the motor.

Solution: We employ Equation 2.11 to solve this problem.

$$\eta_{\text{o}} = \frac{\text{OHP}}{\text{HHP}}$$

We found the output horsepower in the previous example. The hydraulic horsepower is found from Equation 2.10. Thus,

$$\text{HHP} = \frac{\Delta p \times Q}{1714} = \frac{875 \times 8}{1714} = 4.08 \text{ HP (3.04 kW)}$$

Substitution of this value into Equation 2.11 yields

$$\eta_o = \frac{\text{OHP}}{\text{HHP}} = \frac{3.81}{4.08} = 0.934 \ \text{ or } \ 93.4\%$$

Cylinder horsepower output can be calculated from the basic definition of mechanical power:

$$P = \frac{F \times d}{t} \tag{2.12}$$

where F = force
 d = distance
 t = time

However, we are interested here in horsepower, so we divide Equation 2.12 by 550 ft·lb/s to get output horsepower. Thus,

$$\text{OHP} = \frac{F \times d}{550 \times t} \tag{2.13}$$

where F is in pounds
 d is in feet
 t is in seconds

Notice that distance divided by time is velocity; therefore, we can modify Equation 2.13 to get

$$\text{OHP} = \frac{F \times v}{550} \tag{2.13a}$$

EXAMPLE 2.10: A cylinder moves a 5000-lb load at 2 ft/s. Find its output horsepower.

Solution: Using Equation 2.13a, we have

$$\text{OHP} = \frac{F \times v}{550} = \frac{(5000 \ \text{lb})(2 \ \text{ft/s})}{550} = 18.2 \ \text{HP}$$

Interestingly enough, another look at Equation 2.13a brings us back to an already familiar equation. We know that $F = p \times A$ and $v = Q/A$. Substituting these into Equation 2.13a, we have

$$\text{OHP} = \frac{(p \times A)(Q/A)}{550}$$

$$= \frac{p \times Q}{550}$$

When we make the conversions to express $p \times Q$ in ft·lb/s, we arrive at

$$\text{OHP} = \frac{p \times Q}{1714}$$

which is the same as the equation for hydraulic motor output horsepower.

2.3 FLUID POWER SYMBOLS

In this section we take a brief look at the fluid power graphic symbols as presented in standard ISO 1219 from the International Organization for Standardization. The more common symbols are listed in Appendix A.

2.3.1 Rules of Symbol Usage

The following few rules must be followed when using graphic symbols:

a. A graphic symbol represents function only. No attempt is made to describe construction features, location of ports, or anything else peculiar to any specific component design.
b. A circuit drawn with graphic symbols is functional only. It shows the sequence in which components are connected but gives no information as to actual layout or spatial relationships.
c. Valves are drawn in their unactuated positions.
d. The orientation of the symbol on the paper in no way indicates the actual orientation of the component in the actual system and in no way affects the functional meaning of the symbol.

2.3.2 The Language of Graphic Symbols

A first look at a circuit diagram represented by ISO symbols (or any other graphic symbols, for that matter) can be rather bewildering. It falls into the "It's Greek to me" category. An interesting philosophical (or, perhaps, psychological) example concerning graphic symbols may help you overcome some of your bewilderment.

Suppose I asked you to define *tortilla*. You would probably say that it is a flat Mexican bread made of cornmeal (or wheat flour, depending on the type you prefer). You might also tell me that it is used to make tacos or burritos, and maybe some other facts about it.

In reality *tortilla* is simply a group of abstract symbols (called letters, in this case) that, in themselves, have no meaning at all. In fact, even grouped as they are, they have no meaning apart from your own experience. Somewhere along the way, you saw a tortilla and heard somebody call it a tortilla, so it became a tortilla. If you asked a native of Spain to define *tortilla,* he or she would tell you that it is an egg dish similar to an omelet. That person's experience with the word *tortilla* would be completely different from yours.

Just as we put symbols together to make words, we also group words to make sentences. The words in sentences perform different functions. Some are nouns, some are verbs, some are adjectives. The graphic symbols in fluid power circuit diagrams are analogous to the words in sentences. If we put them together properly, they convey important information about the system.

There are relatively few "words" in the graphic symbol vocabulary. There are a few "nouns," some "adjectives," and some "conjunctions." We begin with a circle (noun) and a filled-in triangle (adjective), as shown in Figure 2.4. A circle is used in the symbols for pumps, motors, and gauges but in itself means nothing. In fact, neither symbol means much alone, but if we put them together as shown in Figure 2.5, they make a statement. The triangle tells us that it is not a gauge of any type, so it must be either a pump or a motor. The triangle is shown with its apex pointing out of the circle, meaning the symbol represents an output device (a pump). Since the triangle is solid (filled in), it represents a hydraulic rather than a pneumatic output.

We can also glean some additional information based on what is *not* rather than what *is* included in the symbol. For instance, the single triangle means that there is only one output port. This tells us that we have a pump that can cause fluid flow in only one direction. The absence of a long diagonal arrow tells us there is no way to vary the pump displacement. Consequently, we see that the symbol in Figure 2.5 translates as "a fixed-displacement, single-direction hydraulic pump." However, we do not know if we have a piston, vane, gear, or screw pump.

If we add a long diagonal arrow to this "sentence," as in Figure 2.6, we indicate that the volumetric displacement of the pump can be changed. This symbol means "a variable-displacement, single-direction hydraulic pump." We have no way of knowing whether the pump is a piston pump or a vane pump, but we do know that it is *not* a gear pump, because gear pumps are always fixed displacement.

Thus we can make a multitude of combinations of the basic symbols. You will pick up these concepts quickly with a little practice.

Tying several of these sentences together with the fluid conductor symbols (conjunctions) results in a "paragraph" that functionally defines a complete hydraulic system. Reading such paragraphs should become a matter of routine before you complete this book.

Figure 2.4 A circle and a filled triangle are the basic elements of a hydraulic pump symbol.

Figure 2.5 Combining the elements of Figure 2.4 as shown results in the symbol for a fixed-displacement, single-direction hydraulic pump.

Figure 2.6 Adding the diagonal arrow changes the symbol of Figure 2.5 to represent a variable-displacement, single-direction hydraulic pump.

2.3.3 Circuit Diagrams

Circuits can be broken into two major groupings—open loop and closed loop.

Open Loop   An open-loop circuit is one in which the fluid returning from the system goes into a reservoir. The pump is supplied with fluid from that reservoir. Open-loop circuits may be further categorized as open center or closed center, depending on where the pump is loaded or unloaded during idle periods.

Figure 2.7 shows an open-loop, open-center circuit. In this circuit all fluid returns at low pressure to the reservoir, where it is picked up by the pump to be recirculated. The circuit utilizes a tandem center directional control valve. When the cylinder is not required to move, the valve is centered so that pump output flow passes through the valve at low pressure and returns to the reservoir. The same low-pressure unloading could be achieved by using an open-center directional control valve, but the cylinder could not be stopped and held in position.

Figure 2.7   Open-loop, open-center circuit with a tandem center valve.

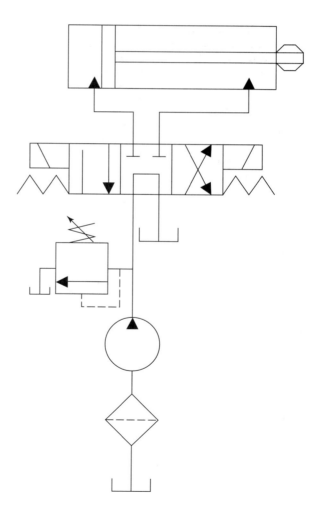

In Figure 2.8, we see an open-loop, closed-center circuit. This circuit uses a closed-center directional control valve. When this valve is centered, the cylinder is stopped and held in place as with the tandem center valve; however, rather than unloading the pump, the closed center "deadheads" the pump. This causes the fixed-displacement pump to operate under its maximum load condition. It works at full pressure and dumps its full output flow across the relief valve at high pressure. This results in energy wastage and high heat generation. It does, however, provide the ability to maintain full system pressure if that is desired.

This closed-center-type operation results when the pressure port of any directional control valve is blocked during circuit idle time. It also occurs when the pressure port of any two-position directional control valve remains connected to a working port after a cylinder has reached the end of its stroke. In short, a closed-center system is any system that forces flow over the relief valve when the circuit is in "idle".

Figure 2.8 Open-loop, closed-center circuit.

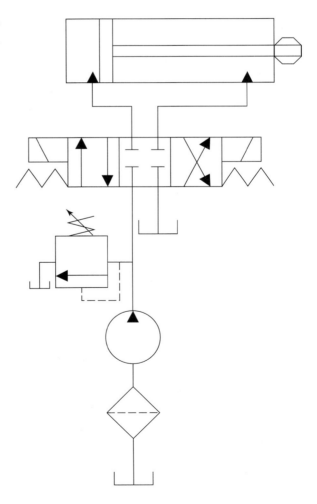

Figure 2.9 A simple diagram of a closed-loop hydraulic system (a hydrostatic transmission).

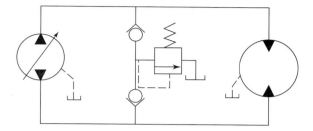

Figure 2.9 shows a *closed-loop* circuit. A closed-loop circuit is one in which the fluid from the power output device (usually a hydraulic motor) does not return to a central reservoir. Instead, it returns to the pump inlet at low pressure to be recirculated through the system. The circuit in Figure 2.9 represents a hydrostatic transmission that utilizes a variable-displacement, bidirectional pump to provide flow for a reversible hydraulic motor. When the pump mechanism is centered, there is no output flow, and the motor will not rotate. Moving the pump mechanism from the center position will cause flow in one direction, resulting in rotation of the motor. Moving the mechanism back across the center position will reverse the flow direction and hence the direction of rotation of the motor.

The circuit, although closed loop, does include a reservoir. Its purpose, however, is simply to collect the case drain flow (the dashed lines from the pump and motor) and from the relief valve (which is used to cushion the motor when it stops) and to ensure that the circuit remains charged with oil.

The symbols and circuits shown as well as the examples discussed in this chapter have been very basic and somewhat simplistic. Their primary purpose has been to refresh your memory. The next chapter will review basic electrical concepts in the same way.

2.4   SUMMARY

This chapter briefly reviewed the basic concepts of fluid power. Little effort was made to describe or categorize the hardware and components that are used in fluid power systems because a basic knowledge of fluid power is assumed. If you need to refresh your memory in this area, some useful references are included at the end of this chapter.

The ability to draw and read circuit diagrams is an absolute necessity for those involved in fluid power system design and maintenance. You should practice this until it becomes as straightforward as reading this sentence. Only when you become fluent in this graphic language can you competently design and maintain systems.

SUGGESTED ADDITIONAL READING

Industrial Hydraulic Technology. 1997. Cleveland, Ohio: Parker-Hannifin Corp.
Norvelle, F. Don. 1995. *Fluid Power Technology,* St. Paul, Minn.: West.
Vickers Industrial Hydraulics Manual. 1993. Rochester Hills, Mich.: Vickers, Inc.

REVIEW PROBLEMS

General

1. A hydraulic cylinder with a 2-in. bore operates with a pressure of 2000 psi. How much force can the cylinder produce?
2. A 3-in. cylinder must produce 5000 lb of force on extension. How much pressure is required?
3. A cylinder has a 3-in. bore and a 2-in. rod. With a system pressure of 3000 psi, what are the maximum extension and retraction forces the cylinder can produce?
4. A cylinder with a $2^1/_2$-in. bore operates with a flow rate of 10 gpm. What is the extension speed of the cylinder?
5. A 3-in. cylinder is required to extend at 4 in./s. What flow rate (gpm) is required?
6. A cylinder with a 2-in. bore and a $1^1/_2$-in. rod operates with a flow rate of 5 gpm. Find the extension and retraction speeds of the cylinder.
7. A cylinder with a 5-cm bore operates with a pressure of 7000 kPa. What is the maximum force it can produce?
8. A 2.5-cm cylinder moves a 5000-N resistance. What pressure is required?
9. A cylinder has a 10-cm bore and a 7-cm rod. What are the extension and retraction forces if the pressure is 10 MPa in both directions?
10. What is the extension velocity of a 5-cm cylinder when the flow rate is 20 lpm?
11. A cylinder with a 10-cm bore and a 7-cm rod operates with a flow rate of 15 lpm. What are its extension and retraction velocities?
12. A 3-in.^{3} (meaning 3-in.^{3}/rev) hydraulic motor operates with a pressure of 1000 psi. The mechanical efficiency of the motor is 0.93. Find the torque it can produce.
13. What size hydraulic motor is required to produce 75 ft·lb of torque at 2000 psi? The mechanical efficiency is 0.95.
14. A 2.5-in.^{3} hydraulic motor operates with a flow rate of 10 gpm. Its volumetric efficiency is 0.82. Find its speed in rpm.
15. What size hydraulic motor will be required to produce 800 rpm at 20 gpm if the volumetric efficiency is 0.88?
16. A 5-cm^3 hydraulic motor is operating at 2000 kPa. It has a mechanical efficiency of 0.95. How much torque can it produce?
17. What size hydraulic motor is required to produce 500 N·m of torque at 3000 kPa? Assume a mechanical efficiency of 0.91.
18. A 3-cm^3 hydraulic motor with a volumetric efficiency of 0.8 operates with a flow rate of 30 lpm. What is its speed in rpm?
19. A cylinder moves a 3000-lb load at 6 in./s. What is its output horsepower?
20. A pump produces 20 gpm at 3000 psi. Find the hydraulic horsepower.
21. A hydraulic motor produces 100 ft·lb of torque at 600 rpm. Find its output horsepower.
22. A cylinder moves a 3-kN load at 20 cm/s. Find its output power in kilowatts.
23. A pump produces 30 lpm at 10 MPa. Find its power output in kilowatts.
24. A hydraulic motor produces 50 N·m of torque at 600 rpm. Find its power output in kilowatts.

Circuit Practice

Note: Always include a relief valve in your circuit drawings. Use standard ISO symbols.

25. Draw a hydraulic circuit to operate a double-acting cylinder. The cylinder cannot be stopped and held except at the ends of its stroke.
26. Draw a hydraulic circuit for operating a double-acting cylinder. The circuit must allow the cylinder to be stopped and held anywhere along its stroke. When the cylinder stops, the pump must operate at low pressure.
27. Draw a circuit to operate a bidirectional hydraulic motor. The motor cannot be stopped while the circuit is operating.
28. Draw a circuit to operate a bidirectional hydraulic motor. The motor can be stopped and held using the directional control valve. The pump must be unloaded when the motor is stopped.
29. Draw a circuit to operate a bidirectional hydraulic motor. Use a three-position valve that will allow the motor to coast to a stop when the valve is centered.
30. Repeat Problem 25, but include meter-in flow control for both directions of cylinder operation.
31. Repeat Problem 25, but include meter-out flow control for both directions of cylinder operation.
32. Draw a circuit for operating two cylinders individually from the same pump.

Computer Exercises

The computer programs required in this section should be written using standard programming language (BASIC, PASCAL, etc.).

33. Write a computer program for calculating the cylinder force when pressure and area are known.
34. Write a computer program for calculating the pressure required to produce a given force from a given cylinder.
35. Write a computer program to determine the size cylinder (bore) to produce a specified force when the pressure is known.
36. Write a computer program for calculating the speed of a hydraulic motor when the size, flow rate, and volumetric efficiency are known.
37. Write a computer program to determine the flow rate required to produce a specified motor speed with a known volumetric efficiency.
38. Write a computer program to calculate the torque produced by a hydraulic motor of a known size and mechanical efficiency.
39. Write a computer program to calculate the pressure required to produce a specified torque when the mechanical efficiency is known.
40. Combine all the preceding programs into a single interactive program that allows you to select the calculations to be done as well as the system of units.

CHAPTER 3

A Review of Basic Electricity

OBJECTIVES

When you have completed this chapter, you will be able to:

- Calculate voltage, current, resistance, and power based on given parameters.
- Recognize the symbols for common electrical sensors and switches.
- Select switches based on circuit power requirements.
- Explain the operation of electrical relays.
- Explain the rules and terminology used in drawing electrical ladder diagrams.
- Draw and interpret simple electrical ladder diagrams.

3.1   INTRODUCTION

The other half of the electrohydraulic marriage is the electrical control circuitry. This circuitry provides command signals to initiate activity as well as feedback information to determine whether (or how well) the commanded function has been completed. In this chapter, we will review some basic electrical concepts (voltage, current, resistance, power, etc.) and introduce switches and sensors. We will also devote some time to introducing ladder diagrams for drawing electrical control circuits.

3.2   BASIC ELECTRICAL CONCEPTS

The electrical equivalents for fluid power parameters are shown in Table 3.1. As in fluid power, certain key relationships of these parameters are important in evaluating electrical circuit performance.

TABLE 3.1

Fluid Power Parameter	Units	Electrical Parameter	Unit
Pressure	psi (kPa)	Voltage	Volt (V)
Flow	gpm (lpm)	Current	Ampere (A)
Power	HP (W)	Power	Watt (W)

3.2.1 Resistance

Resistance is defined as an electrical circuit's opposition to current flow through that circuit. It is designated by the letter R and defined mathematically as

$$R = \frac{V}{I} \tag{3.1}$$

where R = resistance
V = voltage (V)
I = current (A)

The unit of resistance is the ohm, which is designated by the Greek letter omega (Ω). By definition, one ohm is equal to one volt divided by one amp, or $1 \ \Omega = \frac{1 \ V}{1 \ A}$. The resistance of a device is measured with the device disconnected from the circuit and in isolation using an ohmmeter.

Example 3.1: Find the resistance of a coil if a current of 200 mA flows through the coil when a voltage of 9.0 V is applied.

Solution: From Equation 3.1,

$$R = \frac{V}{I} = \frac{9 \ V}{200 \ mA} = 45 \ \Omega$$

An electrical resistor in a circuit is used to limit current flow in the same way that an orifice is used to limit fluid flow in a fluid power circuit. The flow rate through an orifice is determined by the pressure drop across the orifice. The current flow through a resistor is determined by the voltage drop across the resistor. This same concept applies to variable resistors as well as to any electrical component that represents a voltage drop.

Current is measured in series using an ammeter. Voltage (actually voltage drop) is measured in parallel using a voltmeter.

Example 3.2: What value of resistor is required to limit the current through the resistor to 20 mA if the voltage drop across the resistor is 500 mV?

Solution: Again, we employ Equation 3.1:

$$R = \frac{V}{I} = \frac{500 \text{ mV}}{20 \text{ mA}} = 25 \ \Omega$$

3.2.2 Power

Power is defined as the rate of doing work. Electrical work is defined by several useful equations, the most basic of which is:

$$P = V \times I \tag{3.2}$$

where P = power

The units of power from Equation 3.2 are seen to be volt-amperes. This combination of units has been given the designation watt (W), honoring the inventor James Watt for his many contributions to the electrical sciences.

Looking at Equation 3.1, we see that $V = I \times R$, so we can also express power as

$$P = I \times R \times I = I^2 \times R \tag{3.3}$$

Also from Equation 3.1, we have $I = V/R$. Substituting this expression into Equation 3.2 gives us

$$P = V \times \frac{V}{R} = \frac{V^2}{R} \tag{3.4}$$

Many electrical components are rated by the power they produce or consume or by the maximum power they can handle without sustaining damage. For instance, we are all familiar with the 60 W light bulb. We may also read the specification for a switch that states the power limit for the switch in watts.

Example 3.3: A light bulb draws a current of 0.5 A in a 120 VDC circuit. What is the power rating of the bulb?

Solution: From Equation 3.2,

$$P = V \times I = 120 \text{ V} \times 0.5 \text{ A} = 60 \text{ W}$$

Example 3.4: A DC solenoid has a rating of 9.5 W. Determine the maximum current through the solenoid at 12, 24, and 120 VDC.

Solution: Solving Equation 3.2 for current gives us

$$I = \frac{P}{V}$$

For 12 V, this gives

$$I = \frac{9.5 \text{ W}}{12 \text{ V}} = 0.79 \text{ A}$$

For 24 V,

$$I = \frac{9.5 \text{ W}}{24 \text{ V}} = 0.4 \text{ A}$$

For 120 V,

$$I = \frac{9.5 \text{ W}}{120 \text{ V}} = 0.08 \text{ A}$$

Since the power in watts represents the ability of the solenoid to do work, we see that this solenoid can work at a wide range of voltage levels, as long as it is operated at its required power level. If this power level is exceeded at any voltage level, the solenoid is likely to be damaged.

3.2.3 Alternating Current Calculations

The previous examples used direct current (DC) circuits. The equations presented earlier in this section must also apply to alternating current (AC) circuits. Because of the sinusoidal nature of AC power, however, we use the root mean square (rms) values of voltage and current rather than any instantaneous value to calculate the effective values of resistance and power. The rms value of any sinusoidal function is 0.707 of the maximum (peak) value. Therefore, we have

$$V_{rms} = 0.707 \ V_{peak}$$

and
$$I_{rms} = 0.707 \ I_{peak}$$

These are illustrated in Figure 3.1.

In calculating power we can use Equation 3.2 to find that in an AC circuit

$$P_{AC} = V_{rms} \times I_{rms} = (0.707 \ V_{peak}) \times (0.707 \ I_{peak}) = 0.5 \ V_{peak} \times I_{peak}$$

or
$$P_{AC} = 0.5 \ P_{peak} \tag{3.5}$$

Unless otherwise indicated, meters measure rms values.

Figure 3.1 The rms voltage is 0.707 of the peak voltage.

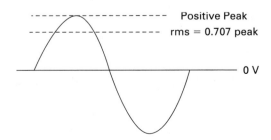

The same concept applies to AC resistance, so Equation 3.1 becomes

$$R_{AC} = \frac{V_{rms}}{I_{rms}} = \frac{0.707 \, V_{peak}}{0.707 \, I_{peak}} = \frac{V_{peak}}{I_{peak}} \tag{3.6}$$

You can see that this is exactly the same as in a DC circuit. Using this in Equations 3.3 and 3.4, we find:

$$P_{AC} = I_{rms}^2 \times R = (0.707 \, I_{peak})^2 \times R = 0.5 \, I_{peak}^2 \times R \tag{3.7}$$

and

$$P_{AC} = \frac{V_{rms}^2}{R} = \frac{(0.707 \, V_{peak})^2}{R} = \frac{0.5 \, V_{peak}^2}{R} \tag{3.8}$$

Example 3.5: An electric heating element draws 12 A from a 120 VAC, 60 Hz circuit. Find:

a. The actual power of the element.
b. The peak instantaneous power of the element.

Solution:

a. The values of voltage and current that we measure in AC circuits are actually the rms values. Therefore, the 120 VAC value is the V_{rms} we need for Equation 3.5. Thus

$$P = V_{rms} \times I_{rms} = (120 \text{ V}) \times (12 \text{ A}) = 1440 \text{ W}$$

b. We can also calculate the peak power from Equation 3.5:

$$P_{peak} = \frac{P_{AC}}{0.5} = \frac{1440 \text{ W}}{0.5} = 2880 \text{ W}$$

We could also use

$$P_{peak} = V_{peak} \times I_{peak}$$

where
$$V_{peak} = \frac{V_{rms}}{0.707} = \frac{120 \text{ V}}{0.707} = 169.7 \text{ V}$$

and
$$I_{peak} = \frac{I_{rms}}{0.707} = \frac{12 \text{ A}}{0.707} = 16.97 \text{ A}$$

so that $P_{peak} = (169.7 \text{ V}) \times (16.97 \text{ A}) = 2880 \text{ W}$

3.3   SYMBOLS FOR COMMON ELECTRICAL SENSORS AND SWITCHES

Like fluid power circuits, electrical circuits and components also have standard symbols that represent the functions, connections, and energy flow paths through the components and circuits. The rules for these symbols are essentially the same as for fluid power graphic symbols. Appendix B includes the more commonly used symbols for these devices.

Of particular importance is the requirement that switches and sensors be drawn in their unactuated positions.

Note: Be aware of a major terminology difference between hydraulic valves and electrical switches. A *valve* that is normally closed (N.C.) has the pressure port blocked in its unactuated position. This means that there is no energy flow through the valve. A *switch* that is normally closed is "on," and there is energy flow through the switch. Conversely, a normally open (N.O.) *valve* has energy flow through it, whereas a normally open *switch* is "off," so there is no energy flow through it. It is becoming common practice in the pneumatics area to use the terms "normally passing" and "normally not passing" instead of "normally open" and "normally closed," respectively. This terminology is more easily related to electrical switching terminology and helps eliminate some confusion.

3.3.1 Switches

In this section we will discuss only a few of the basic electrical sensors and switches. Transducers used for electronic feedback circuits will be discussed in other sections and related to those specific circuits. Sensors and transducers will be discussed in detail in Chapter 8.

It is important to note that the theoretical resistance of any closed switching device is zero. These devices appear to the current flow as a nearly infinitesimal length of wire.

Toggle Switches A *toggle switch* is any switch that uses a lever device to open or close one or more sets of contacts. A common light switch is an example. Toggle switches may be spring loaded or *snap action*. A spring-loaded switch will return to its unactuated position as soon as it is released, much like a spring-returned valve, whereas a snap-action switch utilizes an overcenter mechanism to hold it in any selected position. This switch is analogous to a detented hydraulic valve. The snap-action toggle may also be referred to as a *flip-flop* or a *maintained* switch.

Toggle switches are also described by such terms as "pole" and "throw." The term *pole* defines the number of sets of contacts that are closed when the switch is operated. *Throw* indicates the number of discrete positions in which the switch will close sets of electrical contacts. Figure 3.2 shows the symbols for some common toggle switch configurations. Remember that these symbols show function only; they give no information about the construction features of the switches.

Because toggle switches simply provide a set of contacts, a given switch can be used for either AC or DC switching. There is a limit, however, to the amount of power a switch can handle. The manufacturer's data gives these limits, usually in the form "16 A at 125 V, 8 A at 250 V" or "1 HP at 125/230 VAC/DC". Failure to observe these limits will result in burning up the switch or melting its contacts.

Push-Button Switches *Push-button* switches are actuated by pushing the actuator into the housing. This causes sets of contacts to open or close. Push buttons may be either momentary or maintained. Momentary push buttons return to their unactuated position when they are released, whereas a maintained (or mechanically latched) push button has a latching mechanism to hold it in the selected position. The maintained push button may have

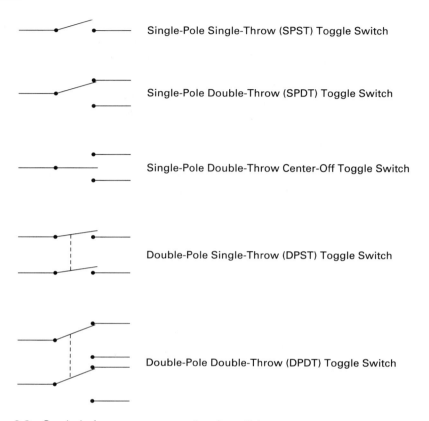

Single-Pole Single-Throw (SPST) Toggle Switch

Single-Pole Double-Throw (SPDT) Toggle Switch

Single-Pole Double-Throw Center-Off Toggle Switch

Double-Pole Single-Throw (DPST) Toggle Switch

Double-Pole Double-Throw (DPDT) Toggle Switch

Figure 3.2 Symbols for some common toggle switches.

a single button that sets with one push and releases with a second (push to set, push to release), or two buttons may be arranged so that one cancels the other. Figure 3.3 shows the symbols for some common push buttons.

If a momentary push button is used, the actuated device will be deenergized as soon as the button is released. If you want the device to remain energized, you must use an electrical latching circuit to keep the current path from being broken when the button is released. As with toggle switches, you must be careful not to exceed the electrical ratings of push-button switches.

Relays *Relays* (also termed *contactors*) are electrically operated switching devices. A common relay (or contactor) application is shown in Figure 3.4. Here, a low-voltage thermostat in the house is used to operate the contactor in the air conditioner. The contactor, in turn, controls the high power necessary to operate the compressor.

Relays generally consist of a magnet coil, one or more moving armatures, and one or more sets of internal contacts. When the coil is energized, the armature is attracted to it. This causes all internal contacts to change state (open to closed, closed to open). Figure 3.5 illustrates this process. In this figure, the relay has one normally open (N.O.) and one normally closed (N.C.) set of contacts. When the coil is not energized (Figure 3.5a),

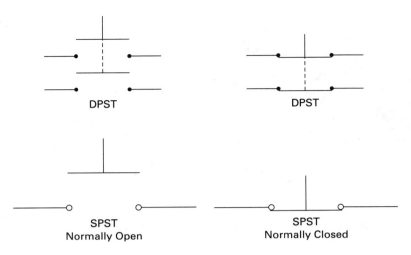

Figure 3.3   Common push-button symbols.

current can flow through the armature from the common contact to the N.C. contact. With the coil energized (Figure 3.5b), the armature is attracted to the iron-magnet core. This opens the N.C. contact and closes the N.O. contact. This allows current to flow from the common contact through the armature to the (now closed) N.O. contact.

Figure 3.6 shows the symbols used to represent relays. The coil is represented by a circle, and the contacts are represented by parallel lines, as shown on the right-hand side

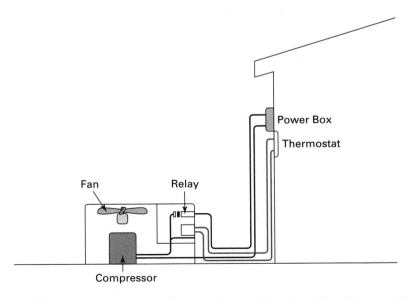

Figure 3.4   The thermostat controls the relay (contactor) coil, which, in turn, controls the compressor and fan.

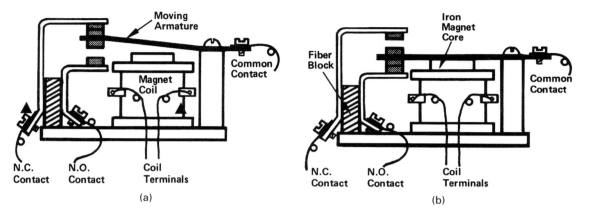

Figure 3.5 Electrical relay with coil deenergized (a); with coil energized (b). (Courtesy of Womack Machine Supply)

of the figure. Notice the designators used in Figure 3.6. The coil is designated by "XCR," where X is the number assigned to the particular relay, and CR indicates a coil relay. (Actually, it does not matter what you call them, as long as you are consistent.) Each set of contacts associated with that relay carries the relay designator and may include a number (or sometimes a letter) denoting that particular set of contacts. In circuit diagrams it is understood that every set of contacts controlled by a relay changes position when the relay is energized, then changes back when it is deenergized.

The electrical rating for relays is more critical than for the switches we discussed previously. The critical component is the magnet coil. The coils can be AC or DC and are rated at one or several voltages. For instance, coils can be rated for 6, 12, 24, 120 VAC or for 6, 12, 24 VDC. Failure to observe these ratings will result in a failed coil.

The contacts in a relay are also rated. These ratings are similar to those used for toggle and limit switches. The contacts simply switch electrical power, so it usually makes no difference whether they are used for AC or DC. As a result, a DC-operated relay may switch either AC or DC power (or both), or an AC relay may switch either AC or DC. As long as the coil and the switched circuits are separate, and the ratings are not exceeded, no problems will occur.

Relays come in many different designs and configurations, so you must always check for the electrical ratings and contact pin arrangements. (The pins are the points to which

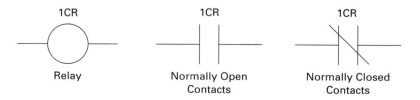

Figure 3.6 Relay and contact symbols.

Figure 3.7 Pin pattern for a standard eight-pin relay. (a) Bottom of relay. (b) Top of base.

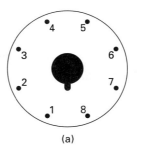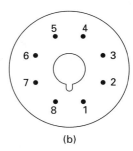

(a)   (b)

you connect the electrical leads for input and output.) This information is shown on the housing for some relays, and in the manufacturers' data sheets for others.

Figure 3.7a shows the pin pattern on a standard eight-pin (or *octal*) relay. On the bottom of the relay, the numbers go clockwise beginning with pin 1 immediately to the left of the alignment key. Figure 3.7b shows the arrangement of the *socket* into which the relay pins are inserted. Notice that it appears to be numbered in the opposite direction of the relay pins, but remember that you are looking at the *bottom* of the relay and the *top* of the socket. Relay design has been standardized so that pin patterns are common for given relay types. Relays may control any number of contacts, so there can be any number of pins. The pins may be arranged in a rectangular rather than circular pattern.

The pin connections to the internal contacts are critical and are standardized within relay types. The connections for the standard eight-pin relay of Figure 3.7 are diagrammed in Figure 3.8, which shows that pins 2 and 7 are used to energize the coil. (*Be careful what you connect to these two pins.*) There are two sets of normally open and two sets of normally closed contacts. Pins 1 and 8 are the commons for these sets. The figure shows that pins 1 and 3 are an N.O. set, as are pins 8 and 6, while 1 and 4 as well as 8 and 5 are N.C.. When using any of the sets of contacts, you should connect the "hot" wires to pins 1 and 8.

Relays may also be classified as holding relays or latching relays. A *holding relay* is wired through its own contacts in such a way that the coil remains energized and holds itself electrically. A *latching relay* is held mechanically.

Figure 3.8 Contact pattern for the pins on a standard eight-pin relay.

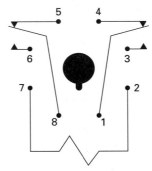

3.3.2 Sensors

The devices to be discussed in this section are somewhat loosely termed *sensors*. In fact, some of them are actually referred to as switches. They are listed under sensors here simply to distinguish them from the manually operated switches in the previous section.

Limit Switches   Any switch that is actuated due to the position of a fluid power component (usually a piston rod or hydraulic motor shaft) or the position of the load is termed a *limit switch*. The actuation of a limit switch provides an electrical signal that causes an appropriate system response.

Limit switches are generally categorized by their actuation mechanism and by their electrical attributes. There are numerous actuation mechanisms, but the most common are mechanical, magnetic, and photoelectric. A mechanically actuated switch is one that requires physical contact to actuate it. A magnetic limit switch (also called a *proximity* switch) is actuated magnetically when the target and the sensing magnet are in the correct positional relationship. Photoelectric switches use the "electric eye" concept to sense the target and actuate the switch. Magnetic and photoelectric switches will be discussed in detail in later sections along with two other types of proximity switches.

Figure 3.9 shows the most commonly used limit switch symbols. These symbols look very much like the toggle switch symbols with the exception of the wedge-shaped actuator symbol on the armature line. If you think of that wedge as being the lever or button that is contacted mechanically by the moving part that is being sensed, you will see that the symbol is very logical and self-explanatory. For a normally open switch, the moving part contacts the lever and pushes it closed. For a normally closed switch, the part pushes the lever away from the electrical contacts and opens the switch. These symbols are used in the *electrical* circuit diagram.

The last symbol in the figure is used in *hydraulic* circuit diagrams. The symbol shows the presence of the limit switch but not whether it is normally open or normally closed.

Pressure Switches   Any switch that responds to pressure or vacuum to open or close a set of contacts is termed a *pressure switch*. A pressure switch that closes a set of contacts when the pressure rises to a preset value is called a *make-on-rise* switch. Functionally, it is normally open. An example of a make-on-rise pressure switch is one that does not allow a system to be put into operation until some preset minimum pressure has been reached. A switch that breaks the contacts when the pressure rises to the preset pressure is termed *break-on-rise*. This type of switch is functionally normally closed. The oil pressure light in an automobile is operated by this type of switch. With the key on but engine off (no oil pressure), the light is on. When the engine starts, the oil pressure rises, breaks the contacts, and turns the light off. If the oil pressure drops below the preset point, the contacts make, and the light comes on again.

Normally, the switch will revert to its original position if the pressure drops slightly below the preset value. This may be undesirable in some circuits, because the pressure could fluctuate around that point. To avoid this, some pressure switches incorporate a feature that keeps the switch from resetting until some substantially lower pressure is reached. The limit switch on an air compressor incorporates this feature. In some switches,

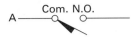

A. SPST (single pole, single throw) N.O. (normallly open) set of contacts. Usually drawn in their "normal" state, which is open, except when there is a reason for showing them in the actuated state.

B. Same contacts as in A, drawn in their actuated state. The side arrow shows they are held closed by a cam before the start of a cycle or during standby periods. However, the position of the solid triangle on the arm, pointing away from the contacts, indicates their normal type and state as N.O.

C. SPST (single pole, single throw) N.C. (normally closed) set of contacts, shown in their "normal"(non-actuated) state. The placement of the solid triangle shows that a cam would open them.

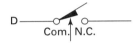

D. Same contacts as in C, drawn in their actuated state. The side arrow shows they are held open by a cam before the start of a cycle or during standby periods. However, the position of the solid triangle on the arm indicates their normal type and state as N.C.

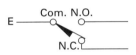

E. SPDT (single pole, double throw) contacts. Sometimes called "transfer" contacts. Placement of the solid triangle on the arm indicates which is the N.O. and which is the N.C. circuit. It indicates also that the contacts are in their "normal" state.

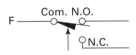

F. Same contacts as in E, drawn in their actuated state. The side arrow shows they are mechanically held in an abnormal state, but the solid triangle on the arm shows their normal state.

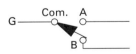

G. SPDT contacts on a maintained type (no spring) switch. A solid triangle on both sides of the arm indicates it must be actuated in both directions. There is no N.O. or N.C. marking on its terminals because it has no "normal" condition.

H. Limit Switch symbol used in hydraulic circuit diagrams.

Note: Abbreviations next to contacts, Com., N.O., and N.C., indicate common, normally open, and normally closed, respectively.

Figure 3.9 Common limit switch symbols.

Figure 3.10 Bourdon tube–type pressure switch. (Courtesy of Womack Machine Supply)

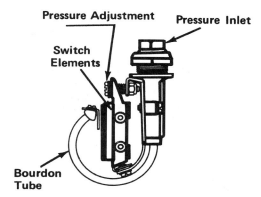

this pressure differential (Δp) is fixed, and in other designs, it is adjustable. Vacuum switches operate in the same way as any other pressure switch except that they respond to vacuums.

The mechanism that responds to the rise or fall of pressure is often a Bourdon tube. This is the same device that is normally used in pressure gauges. Figure 3.10 shows this type of switch. The sensed pressure causes the tube to move sufficiently to open or close the contacts. The actuating mechanism in a pressure switch may also be a bellows, a spring-loaded piston, or a piezoelectric device.

Figure 3.11 shows the symbols for pressure switches. The top two symbols are used in electrical circuit diagrams, and the bottom symbol is used in hydraulic circuit diagrams. Often, a single switch will include both normally open and normally closed contact connectors. Thus the switch can be used as either make-on-rise or break-on-rise, depending on which connector is used. Pressure switches are rated according to maximum pressure, pressure range if adjustable, and the electrical rating of the contacts.

Timers *Timers* (or time delays) are switch mechanisms that keep track of time and then cause one or more sets of electrical contacts to open or close. They can be used to initi-

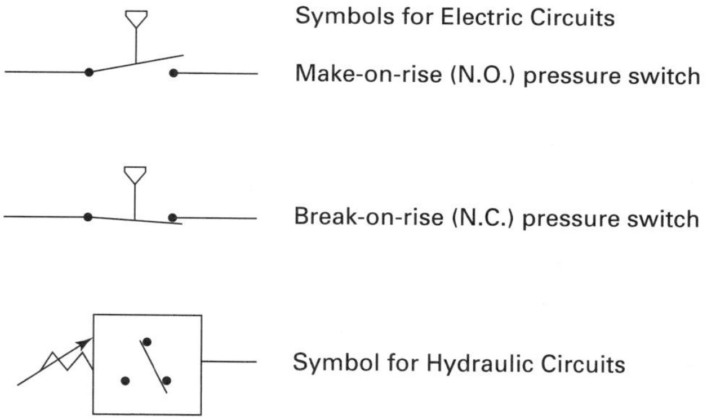

Figure 3.11 Common pressure switch symbols.

Symbols for Electric Circuits

Make-on-rise (N.O.) pressure switch

Break-on-rise (N.C.) pressure switch

Symbol for Hydraulic Circuits

ate or terminate an action or merely change the course of the action. Timers may be electrical, electronic, or mechanical. Pneumatic timers are often used in pneumatic systems. These are devices that must be filled with air before sufficient pressure is available to trigger the next action.

A timer that runs the prescribed time, initiates a single event, then reverts to zero is called a *reset* timer. Such a device must be restarted at the beginning of each event. In contrast, a *program timer* controls a series of events in sequence. The timer on a washing machine is a common example of a program timer.

There are two different modes of operation for timers. The most common is the type that is actuated when power is *applied* to it: Power is applied, and the timer runs out and switches the circuit. This is termed an *on-delay* timer. Once the timer has run out and the switching has occurred, the device remains in the switched condition until power is removed, at which time the device reverts to its initial condition (resets). The timer is restarted by reapplying power. If the power is interrupted before the timer runs out, it resets.

The second type is actuated when power is *removed:* power is removed, and the timer runs out and switches the circuit. This is the type that allows you to leave your car lights on so you can see to unlock the door to your house, then turns the lights off after a short delay. This type of timer is termed *off-delay*. In this case, once the switching has occurred, the device remains in its switched condition until power is reapplied, at which time the timer resets.

The symbols for timers are shown in Figure 3.12. Because timers can either start or stop actions, they can be normally open or normally closed. Timers are rated according to the electrical input for the timing device and the power-handling capability of the contacts.

Sometimes a single timing unit may control several sets of contacts, making it simply a time-delay relay. In this case, the timing unit is shown as a circle, and its contacts are shown in the same way as relay contacts. The timer and all its associated contacts are given a designation such as TMR-1. Contact switching occurs in the same manner as in a relay except for the delay.

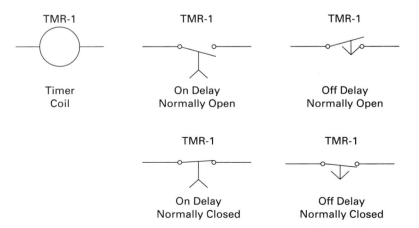

Figure 3.12   Common timer switch symbols.

3.4   ELECTRICAL CIRCUIT DIAGRAMS

There are several ways to draw electrical circuit diagrams, and you may already be familiar with some of them. Such diagrams are usually combinations of standard electrical symbols (such as we discussed in the section on switches and sensors) and pictorial drawings. These diagrams use symbols when it is desirable to show particular arrangements and connections. Pictorial drawings are used when there is nothing peculiar or significant about the component or connections. Figure 3.13 shows a wiring diagram of this type. The relay is shown in standard graphic symbols, and everything else is shown pictorially. Although such wiring diagrams are commonly used, they can be somewhat confusing and difficult to follow. Also, there is no way to standardize these drawings.

Another way to present circuit diagrams is through the use of *ladder diagrams*. These diagrams are described as ladders because the power leads from the *rails* of the ladder, and the *rungs* of the ladder show the series of events required to produce a single output. A simple ladder diagram is shown in Figure 3.14. We will use ladder diagrams throughout this text.

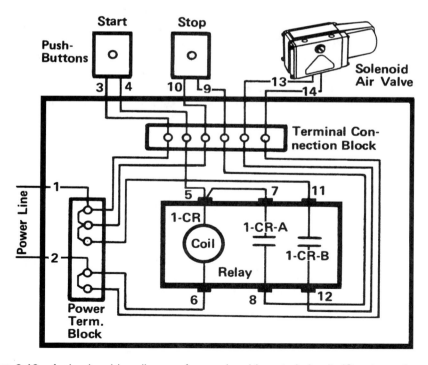

Figure 3.13   A simple wiring diagram for a solenoid control circuit. (Courtesy of Womack Machine Supply)

Figure 3.14 A simple ladder diagram.

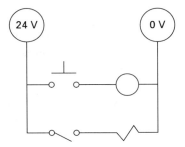

3.4.1 General Rules for Electrical Ladder Diagrams

As presented in the section on sensors and switches, a few basic rules must be observed when drawing electrical diagrams. We look at these first, then at some simple ladders.

a. The ladder may be either vertical or horizontal. In a *vertical* ladder, the "hot" line is on the left, and the ground or neutral is on the right. The first rung is at the top of the ladder with subsequent rungs below it. A *horizontal* ladder has its hot line on the top with the rungs proceeding from left to right.

b. Only graphic symbols are used in ladder diagrams.

c. As far as possible, components should appear in the ladder diagram in the same order in which they occur in the actual system.

d. Relay and timer contacts should be drawn as close as possible to the device that controls them.

e. Relay and timer contacts must carry the same identifier as the relay or timer that controls them.

f. Switches and contacts should be drawn in their "normal" or unactuated state. An exception to this is when a switch is being held in its actuated position. This case is usually shown by drawing the switch in its actuated state and using an arrow to show that it is being held there.

g. With few exceptions, all switching must be done between the hot connection and the output device.

h. Each rung normally defines the sequence of events required to produce a single output. Output devices include motors, solenoids, timers, relays, and the like.

Although it is not a rule, it is common practice to number the rungs for clarity. It is also a good idea to add documentation at the end of the rungs to indicate the function of the rung, where to find the contacts operated by relays and timers, and other information that will help clarify the functioning of the system. It also helps to include the pin numbers on relays and relay contacts for reference during the physical connection of the circuit. Such documentation is very useful to those who will be performing installation or troubleshooting. Even a ladder you draw yourself can be a bit of a mystery after it "gets cold."

3.4.2 Drawing Electrical Ladder Diagrams

The easiest way to talk about drawing ladder diagrams is to draw a few, so let us look at a few simple circuits. First, let us consider a circuit in which a toggle switch is used to turn on an electric motor. We assume that the switch is normally open, so the motor does not start until the switch is closed to complete the circuit. This situation is depicted by the single-rung ladder in Figure 3.15, which is fairly simple and easily understood. When the switch is closed, the motor is turned on—a simple logic sequence.

In our motor circuit it may be desirable to provide a safety circuit that will require that two toggle switches be closed before the motor can start. This arrangement of switches constitutes a series circuit. Figure 3.16 is the vertical ladder diagram of the circuit.

If we wanted to provide two toggle switches arranged so that *either* of them could be used to start the motor, we would have a parallel circuit. This circuit is shown in the ladder of Figure 3.17.

Now let us look at using a momentary push button to start the motor. We diagram this as in Figure 3.18. Such a circuit will work, but it has what may be considered a minor problem—the motor will run only as long as the button is held down. As soon as the button is released, the circuit is broken, and the motor stops. How can we remedy this problem?

One way is to use a mechanically latched push button. We can also do it electrically by using a relay with a holding circuit.

The circuit diagram to do this gets a little more complex, because we must now include the relay and its holding circuit. Let us build this circuit up one step at a time. First,

Figure 3.15 A motor controlled by a toggle switch.

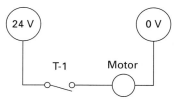

Figure 3.16 Putting the toggle switches in series requires that both be closed to operate the motor (AND logic).

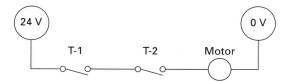

Figure 3.17 In a parallel circuit either switch will operate the motor (OR logic).

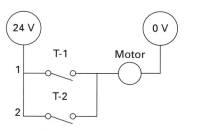

Figure 3.18 A motor controlled by a normally open push button.

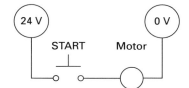

we need to energize the relay. We can do this with a series rung as shown in Figure 3.19a. This rung is similar to Figure 3.17, except here the output device is the relay instead of the motor. We assume a standard eight-pin relay, so pins 2 and 7 are shown as connections. We also still have the same problem—if the push button is released, the relay is deenergized.

We handle this problem by using one of the normally open sets of contacts (1 and 3 in this example) in the relay to provide a holding circuit—a type of bootstrap arrangement—as shown in Figure 3.19b. The sequence we have now is that the button is pushed to energize the relay (1CR), which causes all contacts associated with 1CR to change position. Therefore, contacts 1CR on rung 2 (which are normally open) close. Now, if the push button is released, 1CR remains energized because of the completed circuit through contacts 1CR, its own internal contacts.

Although this is very clever, how does it help us with the motor problem? We now add another rung that contains a second set of normally open contacts [(8 and 6) operated by the relay (1CR)] and the motor (Figure 3.19c). All we have to do to start the motor is close the contacts. That happens automatically as soon as we energize 1CR.

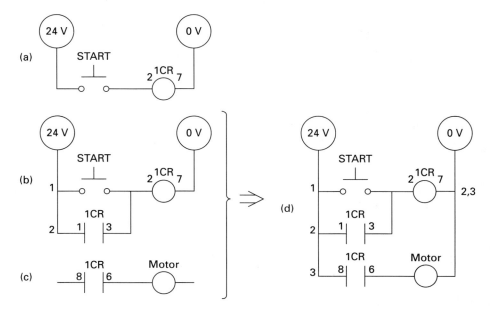

Figure 3.19 The relay circuit with a holding circuit allows the motor to continue to run after the push button has been released.

Figure 3.20 The addition of the STOP push button provides the ability to stop the motor.

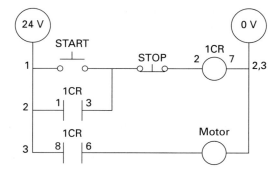

We put the whole circuit together in Figure 3.19d. Talk through the entire sequence of events to ensure that you understand how it works. Also, observe how the rules have been applied. Specifically, note the following:

- The hot line is on the left.
- Sequencing is from top to bottom.
- Switching is on the hot end of the rung.
- There is one output device per rung.

Notice that we have provided documentation for this ladder, even though the ladder is small enough to readily see everything at a glance. The 2,3 on the right-hand end of rung 1 tells us that we can find the contacts controlled by 1CR on rungs 2 and 3 of the ladder. If one of the contacts had been normally closed, this could have been documented by underlining the line number reference.

By now you may have noticed that we have another minor problem with our circuit. How do we stop the motor? Releasing the push button will not do it because of the holding circuit. We must find some way to break the circuit that is energizing the relay. Any ideas?

One way is to put a normally closed push button in the first rung. The placement of this STOP button is critical. It must be placed between contacts 1CR and the relay on rung 1, as shown in Figure 3.20. Now, to stop the motor, we push the STOP button. This breaks the circuit to 1CR and deenergizes the relay coil. All sets of contacts associated with 1CR change state. When 1CR on rung 3 opens, the motor stops. When the STOP button is released, it goes back to its normally closed position, but nothing happens, because the START button and all 1CR contacts are open.

Rungs 1 and 2 of the ladder form a series-parallel circuit. The START and STOP buttons are in series, while 1CR on rung 2 is in parallel with the START button.

3.4.3 Relating Wiring Diagrams to Electrical Ladder Diagrams

You might suspect that a wiring diagram and a ladder diagram that represent the same circuit would tell you the same thing but in different languages. This is exactly the case. Therefore, we should readily be able to relate one to the other. Let's look at an example of this.

Figure 3.13 (used earlier to illustrate a pictorial diagram) shows the wiring diagram for a relay circuit to control a solenoid valve. (We cover solenoid valves in detail later. For now it is sufficient to know that they are electrically operated directional control valves.) Both push buttons are momentary. The START button is normally open, and the STOP button is normally closed. Tracing the wiring, we see that power comes into the circuit via lines 1 and 2 to the power terminal block. Three hot lines are picked off that block. The first goes through the terminal connection block to the START push button (3). The second goes through that block to the STOP button (10). The third goes to contacts 1-CR-B (11).

Pushing the START pushbutton completes the circuit through the coil (5 and 6) and back to the neutral line (2). The jumper from 5 to 7 provides power to one side of contacts 1-CR-A. When the coil is energized, both 1-CR-A and 1-CR-B close. The closing of 1-CR-B completes a circuit from the power terminal block, through 11 to 12, across the terminal connection block to the solenoid (13), out of the solenoid (14), back through the terminal connection block, and back to the neutral line (2) at the power terminal block. In other words, the solenoid is energized.

Because the START button is a momentary push button, what happens when it is released? It pops out and breaks the circuit between 3 and 4. Since that circuit was the one that energized the coil to close the contacts, is the coil now deenergized? No, it is not, because when contacts 1-CR-A closed, a holding circuit was completed from the power terminal block (1), through the normally closed STOP button (10 and 9), through 1-CR-A (8 and 7), to 5, through the coil to 6, and back to the power terminal block (2). This keeps the coil energized and holds 1-CR-B closed. Pushing the STOP button interrupts this circuit and deenergizes the coil, which in turn, deenergizes the solenoid.

If this is the first wiring diagram of this type that you have seen, it may seem a little confusing to you. Actually, this is a very simple circuit. Try to imagine a system that uses several relays to control two or more solenoids each. It can really become a maze. Is there a more orderly way to do things? Yes, there is. That is the function of ladder diagrams.

Figure 3.21 represents the same circuit as Figure 3.13. The relative simplicity of the diagram is immediately apparent. For convenience, the numbering system from the wiring diagram has been repeated on the ladder diagram, so the corresponding points are obvious. Go through the sequence of operation using the ladder diagram. It is easier to see; it is easier to say; and it is much easier to draw. With a little practice, you will be able to read even very complex ladder diagrams quite easily.

Figure 3.21 This ladder diagram represents the wiring diagram shown in Figure 3.13. It is much easier to read and understand.

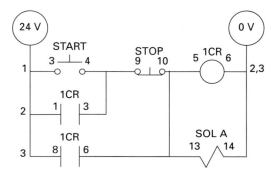

3.4.4 Connecting a Circuit from a Ladder Diagram

The use of ladder diagrams also makes connecting the physical system relatively easy and orderly. First, remember that the hot leg of the ladder represents a single electrical point. That point may be a single terminal or a terminal block with jumpers, but it is still a single electrical point. The same is true of the neutral or ground leg.

Second, progress through the circuit in a logical order, completely connecting one rung at a time. Always start connecting from the neutral line for safety. Keep the physical wiring as simple as possible. It helps to look for common connections that might eliminate extra wires. For instance, in Figure 3.21 it appears that three wires are used to connect 4, 5, and 7. Electrically, these three points are the same, so they can be connected using only two wires.

The "rules" for connecting a circuit from a ladder diagram are summarized as follows:

   a. Work from the neutral to the hot leg.
   b. Connect each rung completely before progressing to the next one.
   c. Simplify the wiring by being alert for common points.
   d. Progress logically down the ladder, connecting each rung in sequence.

3.5   **SUMMARY**

Electrical and electronic controls provide the "brains" to complement the "brawn" of fluid power systems. In this chapter we briefly reviewed the basic concepts of electricity, then extended those concepts to some of the most basic and commonly used sensors and switches used in control circuitry. The terminology of sensors and switches as well as their symbology was introduced. You learned that it is necessary to observe the voltage, current, and power limitations of the sensors and switches when applying them in control circuits.

Electrical relays perform the function of electrically operated switches. They can provide multiple-function control in both simple and complex circuits. We saw in this chapter that on actuation of the coil, all contacts controlled by a relay change state (open contacts close, closed contacts open). We also saw that it is important to adhere to the electrical limitations of both the coil and the contacts when applying relays.

Finally, we introduced the concept of electrical ladder diagrams. These diagrams can be used to show the electrical control circuitry for any kind of motion unit (electrical, electrohydraulic, electropneumatic, or electromechanical). Ladder diagrams are relatively simple to draw and interpret, even though the circuit itself may be very complex. You can make ladder diagrams even easier to use by providing documentation such as line numbers, line reference numbers, and brief explanations of certain control functions.

In electrohydraulic and electropneumatic systems the most basic electrically controlled element is the solenoid-actuated valve. In the next chapter we will discuss these valves in detail, and we will expand the control concepts of this chapter to include the control of basic solenoid control circuits.

REVIEW PROBLEMS

General

1. What is the theoretical resistance of any switching device?
2. When a multimeter is used to measure parameters in a direct current circuit, are the readings peak or rms values?
3. When a multimeter is used to measure parameters in an alternating current circuit, are the readings peak or rms?
4. In switch terminology, what is meant by a *pole*?
5. In switch terminology, what is meant by the term *throw*?
6. Explain the terms *normally open* and *normally closed* as applied to switches.
7. What is the purpose of a limit switch?
8. What is a relay?
9. List the three basic components of a relay.
10. How is a timer different from a relay?
11. How is an ohmmeter used to measure resistance?
12. How is an ammeter used to measure current?
13. How is a voltmeter used to measure voltage drop?
14. In the circuit shown, draw meters in the appropriate locations to make the following measurements:
    a. Voltage drop across switch A.
    b. Voltage drop across switch B.
    c. Voltage drop across the solenoid.
    d. Voltage drop across the entire circuit.
    e. Current through switch A.
    f. Current through the solenoid.
    g. Current through the entire circuit.

Electrical Calculations

15. Find the voltage drop across a 1000 Ω resistor for a current of 5 A.
16. A device has a 500 mV drop and a current of 300 mA. What is the resistance of the device?
17. How much power is required to drive a 120 VAC motor that draws 5 A?
18. How much power is required if 10 A is drawn by a 120 VDC motor? Assuming an efficiency of 0.85, what is the horsepower output of the motor?
19. A 24 VDC solenoid draws 1.6 A. How much power is used? What is the resistance of the solenoid?
20. What is the peak voltage if a meter reads 120 VDC? If it reads 120 VAC?
21. A certain subminiature toggle switch is rated for 6 A at 125 VAC. How much power can it handle?
22. An AC circuit has a peak voltage of 155 V and a peak current of 20 A. What value of current and voltage would you measure with a standard meter? What is the power in the circuit?
23. A certain relay coil is operated on 24 VDC. It has a resistance of 660 Ω. What is its nominal power rating? What is the current through the coil?

24. Draw the standard symbols for the following switches:

   a. SPST toggle
   b. SPDT toggle
   c. DPDT toggle
   d. SPDT center-off toggle
   e. N.O. limit
   f. N.O. proximity
   g. N.C. limit
   h. N.O. pressure
   i. N.C. pressure
   j. N.O. timer (on delay)
   k. N.O. timer (off delay)
   l. N.C. timer (on delay)
   m. N.C. timer (off delay)

Ladder Diagrams

For the following control circuits, draw the appropriate hydraulic circuits and electrical ladder diagrams using standard symbols. Assume standard eight-pin relays. Label each circuit component appropriately, including the relay pin connections. Number each line and provide full documentation. In the hydraulic circuit, cross-reference the ladder diagram and the hydraulic circuit.

25. A single-solenoid DCV is used to cause a double-acting cylinder to cycle automatically one time.
26. A single-solenoid DCV is used to cycle a double-acting cylinder continuously. The operation must be fully automatic.
27. A single-solenoid DCV is used to cycle a double-acting cylinder continuously and automatically for 1 min. At the end of that time, the cylinder stops in the retracted position.
28. A single-solenoid DCV is used to cycle a double-acting cylinder continuously and automatically with a 5-s delay between cycles.
29. A system uses a pump driven by an electric motor to charge a pressure vessel. A relay with a holding circuit is required. A pressure switch is used to maintain the pressure between maximum and minimum levels. A manual shutdown capability is required.
30. A feed grinder can be operated either clockwise or counterclockwise by a hydraulic motor. The appropriate light is illuminated on the operator's panel to indicate the direction of rotation. The control circuit includes (but is not limited to) a master on-off switch, momentary pushbuttons, standard eight-pin relays, and holding circuits for the relays. The hydraulic motor can be stopped without shutting the system down. When the hydraulic motor is stopped, the pump is unloaded.

CHAPTER 4

Solenoid Valves

OBJECTIVES

When you have completed this chapter, you will be able to:

- Recognize the terminology associated with solenoid valves and their related control circuitry.
- Explain the operation of solenoid valves.
- Explain the differences between air-gap and wet-armature solenoids.
- Explain the operation of pilot-operated valves.
- Calculate the time required for a valve spool to shift.
- Draw basic electrohydraulic and electropneumatic circuit diagrams using standard ISO graphic symbols.
- Draw the electrical ladder diagrams depicting the electrical control circuits for the preceding systems and cross-reference the two diagrams.
- Interpret basic Boolean algebra statements.
- Explain the most common causes of solenoid valve failures.

4.1   INTRODUCTION

The solenoid valve is the simplest of the electric and electronic valves used for fluid power control. They were the first electric valves employed in such applications and have been used successfully for almost a century with relatively few major changes to the original concept.

An example of a solenoid valve application is shown in Figure 4.1. The device shown is a three-axis, computer-controlled can crusher used to crush aluminum beverage cans into 1-in. cubes. The three hydraulic cylinders are controlled by solenoid valves, which, in turn, are controlled by the programmable logic controller (PLC). Pushing and holding

Figure 4.1   A three-axis, computer-controlled can crusher.

two push buttons completes the circuit to cause the PLC to energize the first solenoid valve. This causes the first (vertical) cylinder to extend and crush the can axially. When the 1-in. height is reached, a follower on the platen actuates a limit switch, which signals the PLC to deenergize the first solenoid and energize the second one. The second cylinder then extends to crush the can into a rectangle. When the platen reaches the 1-in. position, it actuates a second limit switch, which signals the PLC to deenergize the second solenoid and energize the third. The third cylinder completes the cube. A third limit switch signals the PLC to shift all the valves to retract all three cylinders. During the operation, if either push button is released, the PLC stops the process.

Although this device is hardly a major industrial tool (in fact, it was a student design project), it does illustrate the use of solenoid valves, push buttons, limit switches, PLCs, and other elements that are very commonly used in industrial automation. In this chapter we examine these devices and control circuits in detail. Programmable logic controllers will be covered in Chapter 9.

We will look first at solenoid valve construction, then we will discuss some operating parameters, terminology, and symbols. Next, we will look at some basic solenoid control circuits. We end the chapter with a discussion of causes of solenoid failures and some troubleshooting ideas.

4.2   SOLENOID VALVE CONSTRUCTION

The term *solenoid valve* actually refers to a solenoid-operated or a solenoid-actuated valve. The valve may be operated directly by the solenoid, or the solenoid may be used to actuate a small pilot valve that is used to port fluid to a larger main valve. We will discuss

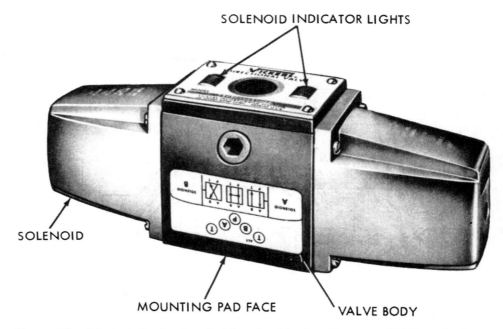

Figure 4.2 A typical direct-acting double-solenoid valve. (Courtesy of Vickers, Inc.)

only directional control valves (including shutoff valves), although there are some pressure and flow applications. The operation of those valves is basically the same as that of the directional valves.

A solenoid valve basically consists of a sliding-spool-type valve section and an actuating solenoid (see Figure 4.2). We will look at these two separate sections in some detail.

4.2.1 Valve Section

The hydraulic valve section is similar to (and in fact, may be identical with) that of a manually or mechanically operated valve except that the spool does not extend through the end of the valve body for connecting to a lever or plunger. Figure 4.3 is a cutaway drawing of a typical direct-acting solenoid-operated directional control valve. It consists of a

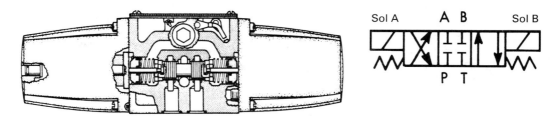

Figure 4.3 Cutaway drawing of a typical direct-acting solenoid-operated directional control valve. (Courtesy of Vickers, Inc.)

valve body, a spool, fluid ports, and return springs. The symbol for the valve illustrated here shows that it is a four-way, three-position valve with an open center spool. A spring on each end ensures that the spool returns to the center position when it is unactuated. The symbol also shows the two solenoids. (There are many other configurations in common use. We will discuss some of those later.)

If neither solenoid is actuated, the springs push the valve spool to its center position. In this position all four ports are blocked, so there is no flow through the valve. Energizing solenoid A pushes the spool to the left. In this position the arrows show that there are two open flow paths. One allows flow from the pressure (P) port to the A port. The other allows flow from the B port to the tank (T) port. Deenergizing solenoid A allows the springs to return the spool to the center position, again stopping all flow.

Energizing solenoid B pushes the spool to the left. Again, the arrows show two open flow paths. This time, flow is from P to B and from A to T. These two end positions allow us to extend and retract a cylinder or run a hydraulic motor in the forward and reverse directions. The closed center position allows us to stop the cylinder anywhere in its stroke or to stop the hydraulic motor.

4.2.2 Solenoid

A solenoid is merely an electromagnet adapted for a particular application. When a current flows through a wire, a magnetic field is set up around the wire. If this wire is wrapped around a ferromagnetic material, that material becomes magnetized when the current flows. (Remember doing this with a nail when you were growing up?) Increasing the number of turns of wire increases the strength of the magnetic field. The magnetic field flows around the coil and through its center as shown in Figure 4.4. If we provide a path through a ferromagnetic material such as iron, we can "shape" the field and concentrate it in a specific space to take advantage of its force capability, as shown in Figure 4.5, where we see the magnetic field flowing through the center of the coil and through the iron structure. The iron structure is commonly referred to as the *C-frame* or the *C-stack*. The coil is normally wound around a plastic spool called the *bobbin*.

So far, we have generated and concentrated the magnetic field, but how do we use it? Into the hollow center of our coil of wire we insert another piece of iron, termed the *T plunger* or simply the plunger (see Figure 4.6). When the coil is energized (current is flowing through it), the concentrated magnetic field draws the plunger into the C-frame

Figure 4.4 If we make a coil of many turns of wire, this magnetic field becomes many times stronger, flowing around the coil and through its center in a doughnut shape. (Courtesy of Detroit Coil Company)

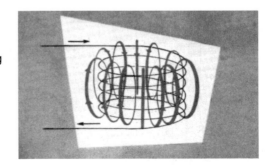

Figure 4.5 Although this magnetic field will flow in air, it flows much more easily through iron or steel—so we add an iron path, or *C stack,* around the coil that concentrates the magnetism where we want it. (Courtesy of Detroit Coil Company)

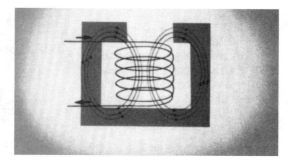

as far as it will go—that is, until physical contact is made between the plunger and some part of the frame.

When the solenoid is not energized and the plunger is resting partially out of the frame, the separation between the plunger and the base of the frame is called the *air gap.* Although the magnetic field will flow through this air gap when the solenoid is energized, the air presents a high resistance, which weakens the magnetic field. As the plunger is pulled in, the air gap becomes smaller. This means that there is less resistance to the magnetic field, so it becomes stronger. Thus, the maximum strength of the magnetic field occurs when the plunger "bottoms out" and the air gap is completely closed. This means that the solenoid has its minimum force when the plunger is out, and reaches its maximum force when the plunger is bottomed out (fully in).

This concept is extremely critical in the design and operation of the solenoids used to operate valves, especially spool-type directional control valves. Due to flow forces, drag, static friction, and contamination, the highest force required to move a spool usually is at the very beginning of the stroke. Unfortunately, this is where the solenoid produces the least force.

The following equation can be used to calculate the force available from a solenoid:

$$F = \frac{1}{2}(NI_0)^2 \frac{\mu_0 A}{X^2} \tag{4.1}$$

where F = initial solenoid force (N)

N = number of turns of wire in the coil

Figure 4.6 If we also add an iron path, known as a *T* or *plunger,* in the center of the coil, the magnetism is concentrated still further. (Courtesy of Detroit Coil Company)

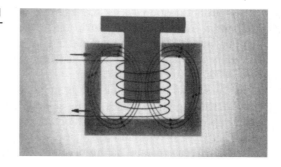

Figure 4.7 As the plunger is pulled into the coil the air gap under the plunger is reduced, making the magnetic field stronger and increasing solenoid force. Thus as the solenoid closes, it becomes more powerful. (Courtesy of Detroit Coil Company)

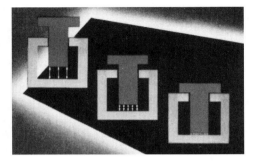

I_0 = initial current (A)

μ_0 = permeability ($4\pi \times 10^{-7}$ H/m in air)

A = plunger cross-sectional area (cm^2)

X = air gap (cm)

As the plunger is pulled into the coil, as shown in Figure 4.7, the air gap (X) gets smaller. As a result, the solenoid force increases.

Example 4.1: Find the initial force capability in pounds for a solenoid that has an 800-turn coil. The plunger is square with 2-cm sides. The initial air gap is 1 cm. The initial (inrush) current is 2 A.

Solution: This is a direct application of Equation 4.1. Therefore:

$$F = \frac{1}{2}(NI_0)^2 \frac{\mu_0 A}{X^2}$$

$$= \frac{1}{2} \frac{(800 \times 2)^2 (4\pi \times 10^{-7})(2)(2)}{1}$$

$$= (6.434 \text{ N})(1 \text{ lb}/4.45 \text{ N})$$

$$= 1.45 \text{ lb}$$

From Equation 4.1 we see that there are some (apparently) obvious ways to increase the initial force capability. For instance, we can increase the number of turns in the coil and/or increase the cross-sectional area of the plunger. However, these steps are applicable only to a certain extent, because there are practical limits to the physical size of the solenoid.

A second possible solution is simply to increase the current. From Equation 4.1, we see that the force increases as the square of the current, so doubling the current means quadrupling the force. Unfortunately, heat generation in a solenoid is also a function of the square of the current (see Figure 4.8). Again, there are practical limitations, particularly in the case of alternating current (AC) solenoids, where the inrush current may be

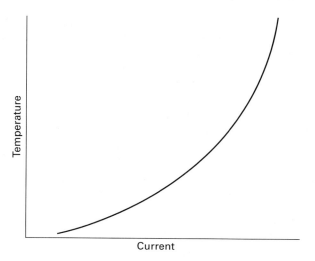

Figure 4.8 The temperature rise is proportional to the square of the current.

4 to 10 times greater than the holding current required when the solenoid is closed. This high inrush current is not a problem with direct current (DC) solenoids, but the heat generation problem still must be considered.

Figure 4.9 shows a typical current versus stroke profile for an AC solenoid. If we project the implications of this graph back into Equation 4.1, we see some rather interesting things. For instance, we start with a high inrush current at the minimum force point, but the current decreases as the plunger is pulled into the core. This seems to imply a rapidly decreasing force capability. In contrast, the air gap is decreasing (in fact, approaching zero). Since the force varies inversely as the square of the air gap, we seem to be moving toward an infinite force.

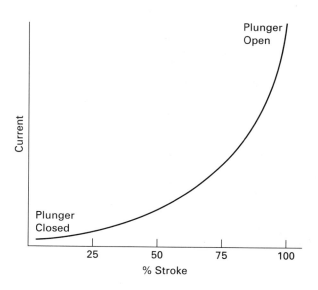

Figure 4.9 The effects of plunger position on AC current.

Another interesting observation is that this high inrush current problem is also a limiting factor in the cycling rate of a solenoid. Consider, for instance, a solenoid with a duty cycle as shown in Figure 4.10. The solenoid is energized 25% of the cycle, so this is termed a "25% duty cycle." The inrush current in this example is 3.5 A. As the plunger moves in, the current reduces until it reaches the design (holding) current, 0.5 A in this case. Although the holding current will generate some heat, solenoids that are designed for continuous duty are capable of dissipating the small amount of heat resulting from the low holding current. The inrush current is the real problem.

Heat is generated by the current, so the area under each energizing pulse represents heat generation in the solenoid. If the heat can be dissipated, there is no problem. However, if the heat generated by continuous cycling of the solenoid exceeds its ability to dissipate the heat, the solenoid will overheat and fail. Obviously, increasing the cycling rate will increase the heat generation rate, whereas reducing the cycling rate will allow more time for heat transfer from the solenoid (Detroit Coil Co., Bulletin 191).

In a DC solenoid, the current (and, consequently, the heat generation rate) stays approximately constant throughout the plunger stroke. This has some significant maintenance and reliability implications, as we will see later.

In addition to the heat generated by energization, the ambient air temperature and the temperature of the operating fluid contribute to the total operating temperature of the solenoid. The maximum total temperature limit of the solenoid is determined by the type of insulation used on the coil. This insulation—usually an epoxy—serves several functions. It insulates the coil electrically, dissipates heat, and seals out moisture, dirt, oil, and corrosive vapors. If its temperature limits are exceeded, the insulation will break down, and the coil will short internally (Parker Hannifin Corp., Bulletin 7325).

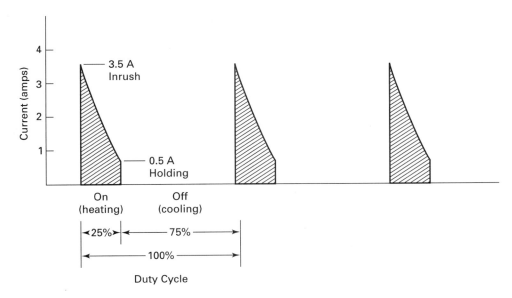

Figure 4.10 Rapid cycling of an AC solenoid, even with a 25% duty cycle, can cause the solenoid to overheat.

TABLE 4.1 Industrial Solenoid Insulation Classifications

Class	Temperature Limit	Type of Insulation
A	220°F	Epoxy or paper
B	266°F	Epoxy
F	311°F	Epoxy
H	365°F	Epoxy

Except for very special high-temperature applications, four classes of insulation are normally used for solenoids. These are listed in Table 4.1.

4.2.3 Electrical Ratings for Solenoids

Solenoids are normally rated by their operating voltage and current. Voltages may be a variety of AC or DC ratings, depending on the solenoid. One major manufacturer, for example, produces solenoids for use on 24, 120, 208, 240, or 480 VAC and 12, 24, and 120 VDC. Other manufacturers' lines may include other voltage levels.

The current, or amperage rating of a solenoid is the maximum amount of current the coil is designed to handle continuously without overheating. The power of a solenoid is determined by its voltage and current. We can calculate this power from Equation 3.2:

$$P = V \times I \qquad (3.2)$$

If the solenoid does not produce sufficient power, it cannot generate the force needed to operate the valve.

In the United States, Canada, and some South American countries, the AC power supplies operate at 60 Hz. In most other countries, 50 Hz is the norm. Most 60 Hz power supplies operate at 115 to 120 V, whereas most 50 Hz systems operate at 100 to 110 V.

Ideally, a solenoid should be designed and fabricated for the specific power supply with which it will be operated. This, in fact, is normally not the case. Most manufacturers rate their solenoids for both 50 and 60 Hz systems. Most of these so-called dual-frequency coils are actually wound for 50 Hz power. When used with 60 Hz power, the coil produces less force than on 50 Hz. For this reason, many solenoids are dual-rated—for example, 120V/60 Hz or 110V/50 Hz. When operated at these specified voltages, the solenoid force is essentially the same with either frequency (Detroit Coil Co., 50- and 60-Cycle Solenoids).

Although it is not recommended by most solenoid manufacturers, a solenoid designed to operate on AC power can be operated on DC power with some limitations. These limitations primarily concern the inrush and holding current characteristics of AC solenoids.

We have already discussed that the AC solenoid has an inrush current that is much higher than its normal operating (holding) current. This high current provides the force necessary to overcome the load imposed on the solenoid during the early part of the stroke, and the low holding current provides a strong force (due to the small air gap) while generating very little heat.

We noted earlier that DC current remains constant. There is no high inrush current or low holding current—just a constant DC current. How is this a problem in the operation of an AC solenoid on DC power? It simply means that we must either compromise or improvise.

If it is necessary to have a high current at the beginning of the stroke to generate a high initial force, there will probably be a problem with heat generation if the solenoid must be held closed. However, if a low heat generation rate is required, then a relatively low current is necessary. This means that the solenoid may not be able to generate the required initial force.

It is obvious, therefore, that there are some applications where an AC solenoid simply will not work on DC power. However, in cases where the solenoid does not need to be held energized, or where the stroke is extremely short, an AC solenoid can usually be operated satisfactorily on DC.

If the solenoid must be held energized, an improvisation can allow the use of an AC solenoid with DC power. The improvisation requires the use of a switch and a resistor. The circuit is arranged so that when the solenoid closes, the resistor is switched into series with the solenoid coil. This allows a high initial coil current that is reduced to a safe holding current when the solenoid closes.

Within limits, DC solenoids can also be operated on AC power. Again, this is not recommended by solenoid manufacturers. The primary problem here is in the difference in the construction of the two types of solenoids. The plunger and C-frame of AC solenoids are normally made of many thin sheets or laminations of iron (see Figure 4.11), and each lamination is coated with insulation. The purpose of this construction is to reduce the heat generation caused by small, stray currents known as *eddy currents*. Because of the insulation, these eddy currents cannot flow from one lamination to another, but the magnetic field is not affected. The result is a stronger, more efficient, and cooler solenoid (Detroit Coil Co., AC Solenoids on DC).

Eddy currents are not a problem in DC circuits, so DC solenoids are often made with solid iron parts. When these solenoids are operated on AC power, eddy current losses occur. These losses may be quite high and, of course, result in high heat generation. Because of this problem, DC solenoids may practically be used on AC power only when a very low current is required or when the solenoid is used in an on-off situation and not held energized.

Another problem that may occur when using AC power on a DC solenoid is solenoid buzz or chatter. In an AC circuit the magnetic field increases and decreases sinu-

Figure 4.11 The C stack and plunger are made of many sheets, or laminations, each of which is laminated. This configuration contains the eddy currents within each lamination. (Courtesy of Detroit Coil Company)

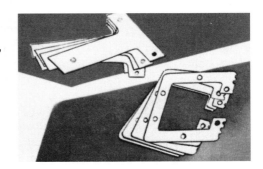

Figure 4.12 To eliminate buzz, and to increase the solenoid holding power, two copper loops, called *shading coils,* are added to the top of the C stack. Current is generated in each of these shading coils, and, most importantly, this generated current lags the applied current. (Courtesy of Detroit Coil Company)

soidally at the frequency of the AC sine wave. When the alternating current is at its positive and at its negative peaks, the magnetic field is at its strongest. As the current passes through zero the magnetic field (and, consequently, the solenoid force) is at its weakest. At this point the load forces on the plunger may lift it from its fully closed position. As the magnetic force increases, the plunger is pulled back in. This motion of the plunger results in solenoid buzz or chatter.

To overcome this problem, most AC solenoids use what are called *shading coils* on the top of the C-frame (see Figure 4.12). As the AC-generated magnetic field builds and collapses, a current is generated in the shading coils. This current lags behind the supply current (see Figure 4.13) and generates a magnetic field of its own. By design, the shading coil magnetic field is at its strongest when the applied current magnetic field is at its weakest. The result, as far as the plunger is concerned, is a near-constant magnetic field that holds it in place and eliminates the buzz problem (Detroit Coil Co., Bulletin 191).

Because DC solenoids do not have a problem with a fluctuating magnetic field, they are normally not manufactured with shading coils. Therefore, when operated on alternating current, they are likely to chatter. Some solenoids are designed to operate on either alternating or direct current. These are usually referred to as AC-DC solenoids.

4.2.4 Transient Suppression

A solenoid is a highly inductive load. A characteristic of any inductor is that when it is first energized, it acts somewhat like an open circuit and causes a delay in the current

Figure 4.13 When the applied current passes through zero, the shading coil current is at its maximum. This low shading coil current provides just enough magnetism to hold the plunger closed when applied current magnetism is at zero, thus eliminating the buzz. (Courtesy of Detroit Coil Company)

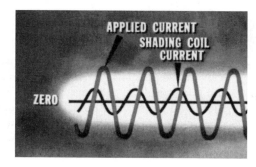

increase before it reaches steady-state operation. This transition to steady state is so rapid that it is of little significance for our applications. However, another characteristic is of more interest to us. When an inductive circuit is disconnected (opened), an extremely high induced voltage spike can result. Therefore, severe voltage spikes or transients can occur when a solenoid is switched. The extent of the disconnect spike can be observed in the violent arcing in the contacts of the relay controlling a solenoid. These high-voltage transients can cause relay contacts to pit or burn out rather quickly. They also represent an extreme hazard in circuits where solid-state components are used. *Transient suppression* is the method used to protect sensitive components from such spikes.

Transient suppression is usually accomplished by an electronic device connected in parallel with the solenoid coil. The device becomes conductive only above a specified voltage that is slightly above the operating voltage of the solenoid. When a voltage spike occurs, the suppressor becomes conductive and allows the spike to flow back into the coil until it dissipates.

Transient suppressors are made for various voltage ranges and are rated by the voltage at which they become conductive. They may be polarized (for use with DC circuits) or nonpolarized (for AC or DC circuits). They normally include semiconductor elements such as diodes, rectifiers, or varistors (Detroit Coil Co., Transient Suppression). A capacitor circuit can also be used for transient suppression. The purpose of the capacitor is to protect the switch from arcing when it opens. It does so by storing electrical energy that would otherwise bridge the gap of the contacts as they open.

4.2.5 Wet-Armature Solenoid

Thus far we have been discussing the traditional solenoid using the C-frame and T-plunger construction. An innovation in solenoid construction represents a major departure from traditional design but remains firmly entrenched in the functional concepts we have already discussed. This innovation is termed the *wet-armature* or *wet-pin* solenoid to distinguish it from the traditional *air-gap* design.

A wet-armature solenoid consists of three functional elements—the tube, the plunger, and the coil (see Figure 4.14). The tube is constructed of magnetically transparent, thin-walled tubing that is closed at both ends. One end may contain a manual override pin (which we will discuss later), and the other end is threaded so that the tube assembly can be screwed into the valve body. The tube assembly contains a cylindrical plunger called the *armature*. The armature fits the tube very loosely and is shorter than the tube so that it has enough movement to stroke the hydraulic valve spool throughout its range of travel.

The coil is of fairly traditional construction, using a U-shaped iron frame to provide a low-resistance magnetic path, and wound copper wire to generate the magnetic field. The coil is encapsulated in an insulating material—usually epoxy—that insulates it electrically and provides protection from the environment. The coil assembly completely surrounds the tube, leaving only the ends of the cylindrical tube assembly exposed (Whitmore). To understand the significance of the wet solenoid unit, we need to see how the solenoid and the valve fit together.

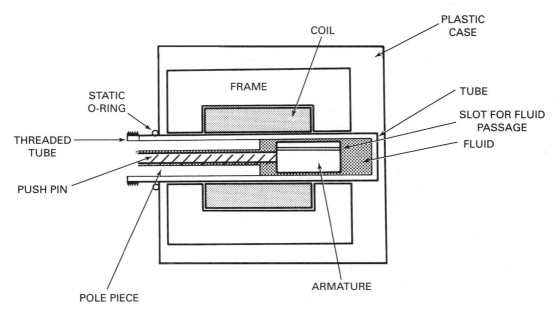

Figure 4.14 Wet-armature solenoid construction. (Courtesy of Vickers, Inc.)

4.3   THE SOLENOID VALVE

We have seen that a solenoid-operated hydraulic valve consists of two main parts—the valve body, which contains the spool, and the solenoid with its plunger. How do these two units work together?

The ends of the valve body are designed to receive the solenoid assembly. If the valve body is to use an air-gap solenoid, the end is flat-faced with a centered hole to the spool cavity with (usually) four bolt holes for attaching the solenoid. A wet-armature valve is also flat-faced, but the centered hole to the spool cavity is larger than for the air-gap solenoid, and it is threaded to receive the threads on the tube assembly. When the tube is screwed into the valve body, the coil assembly is slipped over it, and a "keeper" of some sort (snap ring, screw cap, or a similar device) or bolts are used to secure it.

The assembly now consists of a movable spool in one housing and a movable plunger in another. How do these two movable members "communicate"? They do so by means of a short metal rod (usually stainless steel) termed a *push pin*. The push pin is not attached to either the spool or the plunger but is simply a link between them. When the solenoid is energized, the plunger is pulled in, contacts the push pin, and uses it to push the spool to the desired position. There are no intermediate positions in which standard solenoid actuators are used. The plunger is either fully in or fully out. It never stops in mid-stroke in normal operation.

The primary difference between the air-gap and wet solenoid designs is the way in which the push pin penetrates the valve body. In the air-gap design, the push pin moves through an O-ring seal as it pushes in and out. The dynamic O-ring is designed to prevent loss of oil from the spool cavity into the solenoid cavity (see Figure 4.15). However, each time the pin is pushed out of the spool cavity, the seal allows a very thin film of oil to remain on the pin. As the pin moves back into the spool cavity some of the oil film is scraped off and remains in the solenoid cavity. Under normal operating conditions, the accumulation of oil in the solenoid cavity is one drop for every 3000 to 5000 cycles. Contamination, seal wear, pin misalignment, and other conditions can cause a departure from normal operation and allow fluid to accumulate in the solenoid cavity at a much higher rate. Because this cavity is not designed to contain fluid, it will eventually begin to leak.

In contrast, the wet armature design makes no attempt to keep the fluid from entering the solenoid assembly. In fact, it is designed to operate with the tube assembly full of oil. The only seals provided are two static O-rings—one at the point where the tube assembly screws into the valve body, and one where the manual override pin is inserted (see Figure 4.16). When the valve is operated the first time, the tube fills with oil, which then flows in and out through slots in the armature as the solenoid is alternately energized and deenergized.

Because of its construction, there is seldom a problem with external leakage from a wet-armature solenoid. A second advantage is that the wet-armature solenoid has better

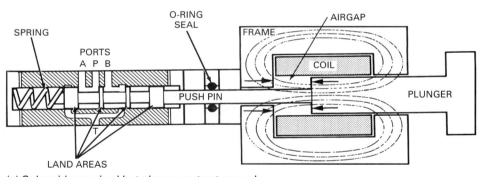

(a) Solenoid energized but plunger not yet moved

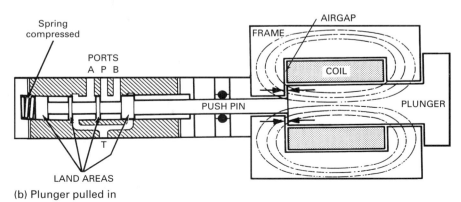

(b) Plunger pulled in

Figure 4.15 Air-gap solenoid operation. (Courtesy of Vickers, Inc.)

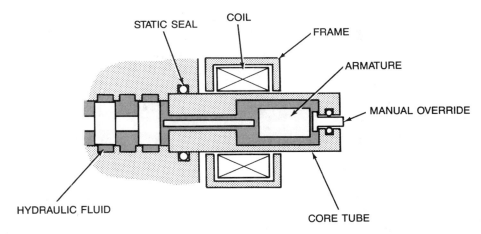

Figure 4.16 A wet-armature solenoid has two static seals, but no dynamic seals. (Courtesy of Vickers, Inc.)

heat dissipation characteristics than the air-gap solenoid because of the fluid in the tube. This characteristic is enhanced during cycling of the valve because the fluid circulation removes heat from the solenoid tube. In fact, up to about two cycles per second, the faster a wet-armature solenoid is cycled, the cooler it will operate (Yeaple 1990). The air-gap solenoid has the advantages of costing less initially and requiring less holding wattage.

4.4 SOLENOID VALVE CONFIGURATIONS

In this section we will examine the various configurations available in the grouping of solenoid valves that are generally known as *direct-acting*. This designation indicates that the valve spool is moved by the force exerted by the solenoid plunger on the push pin. In essence, the solenoid acts directly on the spool.

The valve (actually, the spool) can be of virtually any configuration available for directional control valves—two- or three-position; two-, three-, or four-way; normally open or normally closed; and with any type center position.

There are basically three solenoid arrangements that can be used. We will look at these in both cutaway and symbolic graphics. Figure 4.17 shows a cutaway drawing of a four-way, two-position valve that is solenoid operated and spring returned (or offset). In its unactuated position, there are flow paths between ports P and A, and between ports B and T. When the solenoid is energized, the spool is repositioned so that the flow paths are from P to B, and from A to T. When the solenoid is deenergized, the spring forces the spool back into its original, unactuated position.

The graphic symbol for this valve is also shown in Figure 4.17. In drawings (or interpretations) of solenoid valve symbols, the solenoid is always considered to push the spool. Likewise, the spring is considered to push the spool to its unactuated position. Therefore, we see that this symbol exactly represents the operation of the valve. When

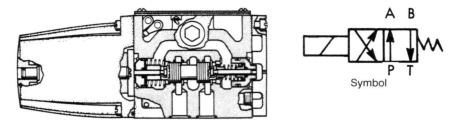

Figure 4.17 Cutaway drawing of a single solenoid valve. (Courtesy of Vickers, Inc.)

the solenoid is not energized, paths P–A and B–T are open. Energizing the solenoid results in the spool's shifting to open paths P–B and A–T. As long as the solenoid is energized, the spool remains in that position. When the solenoid is deenergized, the spring pushes the spool back to the unactuated position.

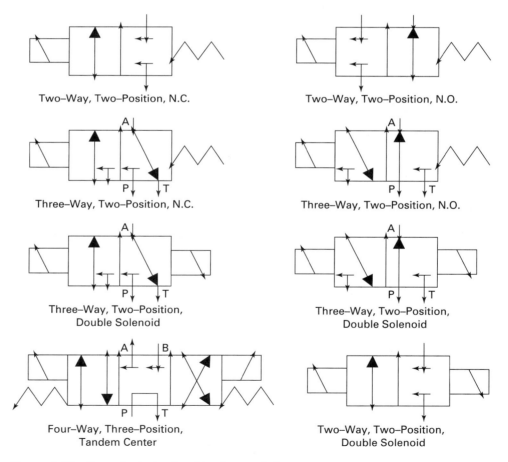

Figure 4.18 Some typical solenoid valve symbols.

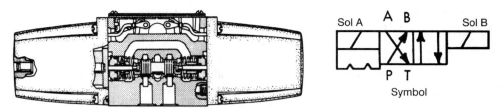

Figure 4.19 Two-position double-solenoid directional control valve (showing detent). (Courtesy of Vickers, Inc.)

The functioning of this valve could be reversed simply by exchanging the positions of the solenoid and the spring. This would provide paths P–B and A–T when unactuated, and P–A, and B–T when actuated.

From this discussion, you can visualize numerous other possible configurations of these solenoid-operated, spring-returned valves. Figure 4.18 shows the graphic symbols for some of the more common arrangements.

A variation on the valve in Figure 4.17 is the use of a second solenoid in place of the spring (see Figure 4.19). In this arrangement, there is no "normal" or "unactuated" position. When solenoid A is actuated, the spool is pushed to the right, opening flow paths P–A and B–T. If solenoid A is deenergized, the spool will remain in that right-hand position. There is nothing to cause it to move to the left unless solenoid B is energized. Then the spool will move to the left end and remain there until solenoid A is again energized. To prevent spool drift when neither solenoid is energized, a detented spool is sometimes used. This configuration utilizes a spring-loaded ball that sits in a groove or notch in the spool to hold it in place. The spring exerts very little force on the ball, so the solenoid can easily shift the spool out of the detented position. When using any double-solenoid configuration, you must ensure that your control circuit cannot cause both solenoids to be energized at the same time. Otherwise, improper operation or solenoid failure will occur.

Solenoids can also be used to operate three-position valves (see Figure 4.20). In these valves there are *always* (a dangerous word) two solenoids and two springs. The center position (which can be any configuration) is considered the unactuated position. The function of the two springs is to return the spool to the center position when neither solenoid is energized.

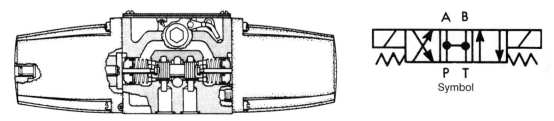

Figure 4.20 Three-position double-solenoid control valve. (Courtesy of Vickers, Inc.)

4.5   PILOT-OPERATED SOLENOID VALVES

Thus far we have discussed only direct-acting valves. These valves are relatively simple in construction and are usually relatively small physically. Their small size is dictated by the force capability available in standard solenoids.

In our previous discussion of the implications of Equation 4.1 we saw that there are practical limitations to the amount of force that can be generated with standard solenoid designs. A force capability of somewhere around 15 lb can be considered to be on the high end of the scale. A force of 8 to 10 lb may be required to move the spool (considering spring resistance, static and dynamic friction, flow forces, and other factors). Thus there is little force left to overcome the effects of contamination, sludge formation, and any other problems that might impede the spool movement. The larger the spool, the larger the force required to move it. For this reason, you will seldom see a direct-acting solenoid valve with porting larger than 3/8 in. In larger valves, standard solenoids cannot reliably produce the force required.

Does this mean that we are stuck with this relatively small size limitation for solenoid valve applications? Not at all. The force problem is actually solved rather simply. Before we look at the solution, though, let us look at a concept.

Figure 4.21 is a partial circuit diagram showing a four-way, two-position, solenoid-operated, spring-returned directional control valve used to operate a double-acting cylinder. In its unactuated position (as shown), pump flow passes from the pressure (P) port to the A port and into the cap end of the cylinder. This forces the piston to the right. The fluid in the rod end of the cylinder is forced out, flowing through the valve from the B port to the T port and back to the system reservoir (not shown). Energizing the solenoid moves the valve spool to the right. This reverses the flow directions through the valve and causes the cylinder to move to the left.

Now, let us remove the cylinder and install in its place a large valve (Figure 4.22). We will connect the lines from the solenoid valve to the ends of the spool cavity in the same way we did with the cylinder. If we operate the solenoid valve as before, we can cause the spool of this large valve to move just as the piston of the cylinder moved. This time, however, instead of extending and retracting a rod, we are sliding the valve spool

Figure 4.21 A cylinder operated by a direct-acting solenoid valve.

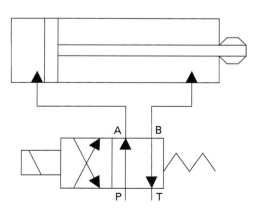

Figure 4.22 The operation of the main spool in a pilot-operated valve is similar to the operation of a hydraulic cylinder.

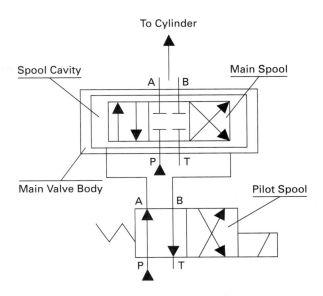

to open and close the flow paths through the large valve. We are simply using fluid power to accomplish what the solenoid alone could not do.

In practical applications the solenoid valve is stacked on top of the large (main) valve to make a single unit (see Figure 4.23). This configuration is often termed *piggy-backing*, and the result is commonly called a *pilot-operated* directional control valve. The small, direct-acting solenoid valve is called the *pilot valve*, and the large valve is usually

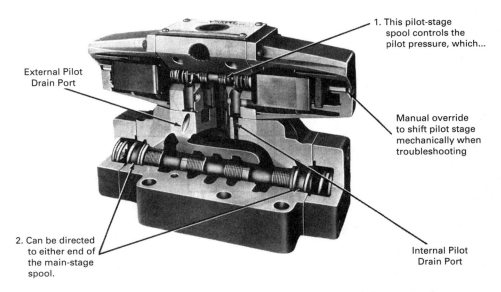

Figure 4.23 Cutaway of a piggyback configuration. (Courtesy of Vickers, Inc.)

referred to as the *main spool*. Because the main spool operates with hydraulic pressure instead of a solenoid, the size restrictions (which caused the force problem when the direct-acting valve was used) are virtually eliminated.

The pilot valve is normally one of three types—four-way, two-position, solenoid-operated, spring-returned; four-way, two-position, double-solenoid-operated; or four-way, three-position, solenoid-operated, spring-centered with a float center. The main spool, however, can be of virtually any configuration. The system designer or user has essentially the same flexibility as with direct-acting valves.

The graphic symbology for pilot-operated valves, like other symbols, illustrates valve function. In this case, however, there is considerably less rigidity than with most components. Look at Figure 4.24, for example. The symbol shows that the main valve has a four-way, three-position, spring-centered, closed-center spool. The pilot valve uses a four-way, three-position, double-solenoid-operated spring-centered, closed-center spool.

The complete (and correct) symbol for this valve must include an exact functional representation of both the pilot and main sections with the appropriate connecting lines as shown. Although this is the *correct* way, it is very seldom used. In common practice, a simplified version is used (also shown in Figure 4.24). It is in this simplified symbol that there is some flexibility. Basically, the symbol must describe the valve function. For a pilot-operated valve, the symbol must describe the main spool, the solenoid and pilot actuators, and the action of the main spool when there is no solenoid actuation. Conceivably, this may be done in different ways for the same valve. If the symbol accurately describes the valve function, the symbol is acceptable.

Figure 4.24 Symbols representing the valve shown in Figure 4.23.

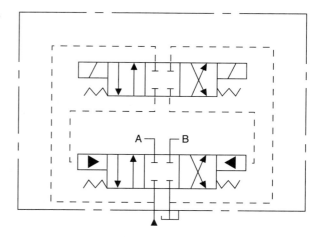

(a) Complete Symbol

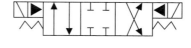

(b) Simplified Symbol

The pilot pressure (that is, the pressurized fluid the pilot valve directs to operate the main spool) may be the same as the system operating pressure. If this is the case, the pressure is transmitted via internal porting within the valve body from the main pressure port to the pressure port of the pilot valve, and is termed *internal pilot pressure*. The symbol for this valve is shown in Figure 4.25.

In some cases, however, when internal piloting is not practical, for example, when the system pressure fluctuates widely and would make the pilot pressure unpredictable or unreliable, or when the system pressure is so high that it could cause very rapid acceleration of the main spool and lead to a water hammer effect in the system. (In most cases, a pilot pressure of 300 to 400 psi (2 to 3 MPa) is preferred.)

In either of these cases it may be desirable to provide the pilot pressure from a more stable or controlled source separate from the system pressure to the main valve, that is, an *external pilot pressure*. It is physically ported into the pilot valve through an external pressure port (see Figure 4.26). In this situation the internal pilot pressure passageway will be blocked.

The same options are available for the drain (tank) lines for the pilot valve. An *internally* drained valve utilizes an internal passage to route the draining fluid to the tank port of the main valve (see Figure 4.27). If there is a danger of excessive pressure in the tank line, or if there are pressure surges in the tank line that may cause the valve to shift in an undesirable way, the pilot spool may be drained through an external drain port that is connected directly to the tank. Thus this is an *externally* drained valve (see Figure 4.28).

Figure 4.25 Symbols for an internally piloted valve.

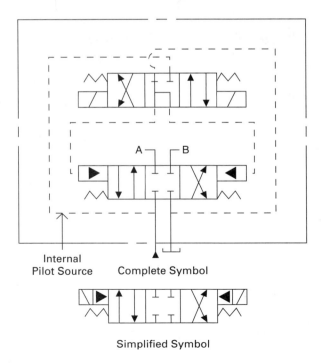

Internal
Pilot Source   Complete Symbol

Simplified Symbol

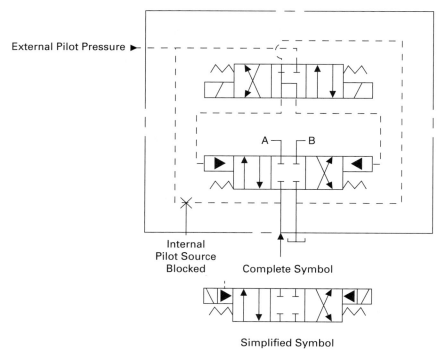

External Pilot Pressure ▶

Internal
Pilot Source
Blocked

Complete Symbol

Simplified Symbol

Figure 4.26   Symbols for an externally piloted valve.

Figure 4.27   Symbols for an
internally drained valve.

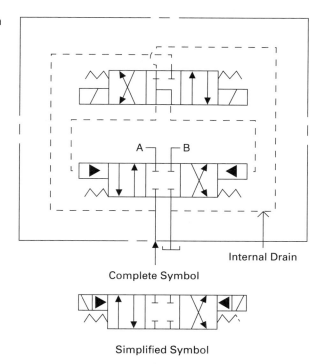

Internal Drain

Complete Symbol

Simplified Symbol

Figure 4.28 Symbols for an externally drained valve.

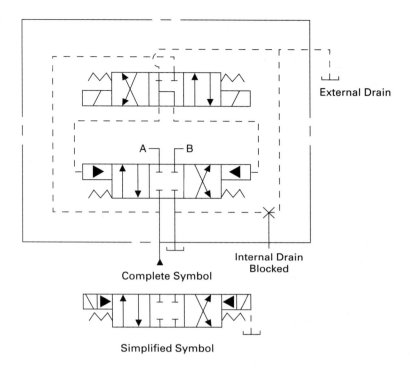

External Drain

A — ⌐ B

Internal Drain
Blocked

Complete Symbol

Simplified Symbol

4.6   VALVE SHIFTING TIME

In calculations of the shifting time of a pilot-operated valve, both the pilot valve speed and the main spool speed must be considered. These two speeds can be calculated separately, then combined to find the total operating time.

4.6.1 Pilot Spool

In order to calculate the shifting time of the pilot spool, we need some information about both the solenoid and the pilot spool. Earlier, we used Equation 4.1 to calculate the force available from the solenoid based on the current and the plunger position. If we have information concerning the initial and final solenoid forces and the plunger stroke, we can approximate the solenoid force at any plunger position using the following much simpler equation:

$$F_x = F_i + Bx \tag{4.2}$$

where F_x = solenoid force at any point x (lb, N)
   F_i = initial solenoid force at maximum displacement (lb, N)
   B = force gradient (lb/in., N/cm)
   x = displacement from the unactuated position (in, cm)

This equation assumes a linear force gradient. That assumption is not accurate, but over a short stroke, it is approximately correct.

Example 4.2: A solenoid has a 0.35-in. (0.89-cm) stroke. The initial solenoid force is 1.2 lb (5.3 N), and the final force is 8.0 lb (35.6 N). Find the force when the plunger has traveled 0.15 in. (0.38 cm).

Solution: We will use Equation 4.2 to find the force, but first we must find the value of B, the force gradient.

$$B = \frac{\text{final force } - \text{ initial force}}{\text{total stroke}} \tag{4.3}$$

$$= \frac{F_f - F_i}{S}$$

$$= \frac{(8 - 1.2)\text{lb}}{0.35 \text{ in.}}$$

$$B = 19.43 \text{ lb/in. (34.0 N/cm)}$$

Now, using Equation 4.2, we get

$$F_x = F_i + Bx$$

$$= 1.2 \text{ lb} + (19.43 \text{ lb/in.})(0.15 \text{ in.})$$

$$= 1.2 \text{ lb} + 2.9 \text{ lb}$$

$$F_x = 4.1 \text{ lb (18.2 N)}$$

We could also use Equation 4.2 to plot a linear approximation of the force/air-gap curve of the solenoid (see Figure 4.29). We have already seen from Equation 4.1 that the force is not a linear function, but because of the short plunger stroke, the linear approximation is reasonable.

Figure 4.29 The actual solenoid force curve can be approximated by a straight line with slope B over a short distance.

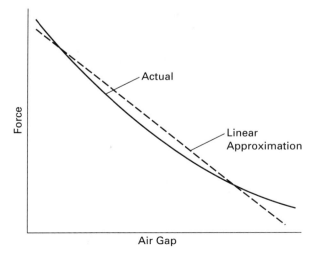

The equation for calculating the spool shifting time includes factors from the force/air-gap curve. As with the solenoid force calculation, we can approximate the spool shifting time by making a simplifying assumption, namely, neglecting viscous damping forces. This assumption is usually valid because of the short stroke and relatively loose fit of the spool in the valve body. The relatively low spool velocity through a low-viscosity oil also supports this assumption. Thus we can approximate the time to reach any displacement x using the following equation:

$$t = \sqrt{\frac{m}{B}} \ln \left[\frac{x + r + \sqrt{2\,rx + x^2}}{r} \right] \tag{4.4}$$

where t = shifting time (s)
  m = spool mass (lb·s²/in., N·s²/cm)
  $r = F_i/B$ (initial force divided by force gradient) (in., cm)
  x = displacement from initial position (in., cm)
  B = force gradient (lb/in., N/cm)

Example 4.3: A direct-acting solenoid valve has the following characteristics. Find the shifting time of the valve.

Spool mass = 0.00018 lb·s²/in. (0.0315 kg = 0.000315 N·s²/cm)

Spool stroke = 0.15 in. (0.381 cm)

Solenoid force

  initial = 0.85 lb (3.78 N)

  final = 7.5 lb (33.36 N)

Solution: To use Equation 4.4 we first need to find the values of B and r:

$$B = \frac{F_f - F_i}{S} = \frac{(7.5 - 0.85)\text{lb}}{0.15 \text{ in.}} = 44.33 \text{ lb/in. (77.64 N/cm)}$$

$$r = \frac{F_i}{B} = \frac{0.85 \text{ lb}}{44.33 \text{ lb/in.}} = 0.019 \text{ in. (0.0483 cm)} \tag{4.5}$$

The shifting time, then, is

$$t = \sqrt{\frac{0.00018 \text{ lb·s}^2/\text{in.}}{44.33 \text{ lb/in.}}} \ln \left[\frac{0.15 \text{ in.} + 0.019 \text{ in.} + \sqrt{(2)(0.019 \text{ in.})(0.15 \text{ in.}) + (0.15 \text{ in.})^2}}{r} \right]$$

$$= 0.0031 \text{ s}$$

$$= 3.1 \text{ ms}$$

The shifting time calculated in the preceding example is somewhat less than we would find if we actually timed the valve. In general, the shifting time of an AC solenoid valve should be from 6 to no more than 16 ms. If the valve has not shifted within 16 ms (one full cycle of a 60 Hz signal), chances that the spool will shift are greatly reduced (Whitmore).

Solenoid valves operating on DC power may have a slightly longer shifting time. The time required for the valve spring to return the spool ranges from 22 to 45 ms.

4.6.2 Main Spool

A pilot-operated valve has an additional dynamic system; therefore, the shifting time is significantly longer because of the larger mass of the main spool as well as the hydrodynamic forces resisting the spool movement. These forces, according to Yeaple (1990), result from pressure in the return line to the pilot section (which effectively generates a back pressure that opposes the spool movement), viscous damping in the main valve section, and axial and radial force components due to the incoming fluid stream as the main spool begins to move.

Most of these factors are fixed by the design of the main and pilot sections and by the passageways between these two sections. The user can still exercise some control over the main spool velocity (hence, shifting time) by controlling two factors: the pilot inlet pressure and the flow rates between the two sections.

Recall that the spool in the main valve is shifted in the same manner in which the piston in a hydraulic cylinder is moved—pressure and input flow at one end, exhaust flow from the opposite end. Also recall that, assuming sufficient pressure to overcome the resistances, the speed at which the cylinder will move is determined by the incoming flow rate and can be calculated from the equation

$$v = \frac{Q}{A} \tag{4.6}$$

where
v = velocity
Q = flow rate
A = effective area

The velocity of the main spool can also be calculated from this equation. Consequently, since the flow rate determines the velocity, controlling the flow rate controls the spool shifting time. If the spool shifts too rapidly, it can be slowed by reducing the flow rate from the pilot section into the cavity at the end of the main spool. This reduction in flow rate can be accomplished in two ways.

One method adapts a technique for controlling cylinder speed—the use of a meter-out flow control valve. In the pilot-operated valve application, a ported plate containing the flow control mechanism is sandwiched between the pilot valve body and the main valve body. This section—called a *choke control*—normally contains two needle valves to perform the flow control function, and two check valves to provide free flow past the needle valves in the direction in which control is not desired. This arrangement allows independent control in each direction (see Figure 4.30).

A second method for slowing the shifting time is to reduce the pilot pressure. The parameter can be used effectively because the flow rate through the pilot valve responds to the pressure drop across the valve. This flow rate can be estimated by using a form of the orifice equation, specifically:

$$Q = CA \sqrt{\frac{2 \, \Delta p}{\rho}} \tag{4.7}$$

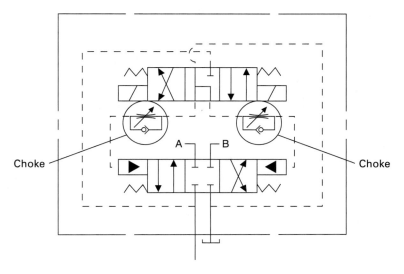

Figure 4.30 Pilot chokes can be used to control the shifting speed of the main valve.

where   C = discharge coefficient (0.5 to 0.7)
         A = effective orifice area
         Δp = pressure drop across the pilot valve
         ρ = fluid density

The pressure drop across the pilot section is a function of the pilot pressure (on the inlet port) caused by drag, flow forces, and spring resistance in the main spool as well as the back pressure in the passageway that directs the flow from the opposite end of the main spool. Thus, decreasing the pilot pressure causes a reduction in the pressure drop with a consequent reduction in pilot flow and a slowing of the main spool speed. Conversely, an increase in the pilot pressure will result in an increase in the main spool velocity.

Example 4.4: Consider a pilot-operated directional control valve that uses the valve of Example 4.3 for its pilot section. Other parameters of the valve are as follows:

Pilot pressure: p = 120 psi (828 kPa)

Back pressure: p_{back} = 50 psi (345 kPa)

Flow coefficient: C = 0.65

Effective orifice area: A = 0.06 in.^{2} (0.387 cm^2)

Density: ρ = 9 × 10^{-5} lb·s^2/in.^{4} (9.62 × 10^{-6} N·s^2/cm^4)

Main spool area: A_{main} = 3.2 in.^{2} (20.65 cm^2)

Main spool stroke: S = 2.2 in. (5.59 cm)

Solution: The total valve shifting time will be the sum of the time to shift the pilot spool plus the time to shift the main spool. (In reality, the main spool will begin to move before the pilot spool completes its stroke, but we will consider these actions to be separate and sequential.) We have already found that the pilot spool shifts in 3.4 ms. Now we need to calculate the main spool shifting time.

First, we must find the flow rate entering the main spool chamber. We use Equation 4.7 for this:

$$Q = CA\sqrt{\frac{2\,\Delta p}{\rho}}$$

We find the pressure drop by subtracting the back pressure from the pilot pressure. Therefore,

$$\Delta p = p - p_{\text{back}} = 120 - 50 \text{ psi} = 70 \text{ psi (483 kPa)}$$

This allows us to calculate the flow rate:

$$Q = (0.65)(0.06 \text{ in.}^2)\sqrt{\frac{(2)(70 \text{ lb/in.}^2)}{9 \times 10^{-5} \text{ lb·s}^2/\text{in.}^4}}$$

$$= 48.6 \text{ in.}^3/\text{s } (796.6 \text{ cm}^3/\text{s})$$

Now we can find the velocity of the main spool from Equation 4.6:

$$v = \frac{Q}{A_{\text{main}}}$$

$$= \frac{48.6 \text{ in.}^3/\text{s}}{3.2 \text{ in.}^2} = 15.2 \text{ in./s } (249.1 \text{ cm}^3/\text{s})$$

The shifting time for the main spool is then

$$t_m = S/v$$

$$= \frac{2.2 \text{ in.}}{15.2 \text{ in./s}}$$

$$= 0.1447 \text{ s} = 147.7 \text{ ms}$$

To this result we add the pilot spool time of 3.4 ms, to get a total shifting time of

$$t_t = t_p + t_m$$

$$= 3.4 + 144.7 = 148.1 \text{ ms}$$

This is a fairly typical (if there is such a thing) shifting time. A reasonable rule of thumb is to approximate the total shifting time of a pilot-operated directional control valve at around 200 ms.

4.7   VALVE MOUNTING

Directional control valves—manual or solenoid and direct-acting or pilot—can be joined to the system by one of two methods. In the first of these methods, the valve itself has threaded ports to accept either threaded fittings (to which a pipe, hose, or tube is subsequently attached) or a pipe or pipe nipple. This design, referred to as *in-line* mounting, is very commonly found on mobile equipment but is much less common in industrial applications.

Most industrial solenoid valves use a *subplate* mount. In this design, the valve itself has no threaded ports. Rather, its ports are drilled into the base of the valve body. The surface containing these ports must be flat within 0.0005 in. and smooth within 45 μin. Recesses are machined around each valve port so that an O-ring seal can be used between the valve and the subplate.

The subplate itself is a metal block containing ports for mating with the valve body, appropriate internal passageways, and threaded ports designed to accept the system conduits. The mating surface of the subplate is also finely machined to provide proper mating with the valve base.

The design of the subplates has been standardized, and there are relatively few from which to choose. Actually, the subplate is chosen by default, because the size and type of valve chosen dictates the subplate to be used, leaving the user to specify only the location of the threaded ports (bottom or sides) and the thread for the ports. The international standard for subplates is ISO 4401.

The standard configurations for subplates are shown in Figure 4.31. The designator (ISO 4401-03, -05, etc.) specifies both the port pattern and dimension. In the case of subplates for pilot-operated valves, it may also be necessary to specify where plugs or orifices are required for the pilot pressure and drain ports.

This subplate mounting method allows for very rapid valve changes. The valve is attached to the subplate by four bolts, so a valve removal involves simply disconnecting the solenoid leads and removing the four mount bolts. To replace the valve, ensure that the O-rings are placed properly, set the valve on the base and insert and torque the bolts, then connect the solenoid leads. The use of connectors rather than soldered leads will make this task much simpler. Numerous options are available.

Now, let us briefly review the subject of port threads, an area with which you should already be familiar. The purpose of the threads is to mechanically attach the conduits to the subplate. It is highly desirable that this attachment be leak free, also. The mechanical attachment may be accomplished by threading the conduit or appropriate fitting directly into the subplate or by using an external clamping device known as a *four-bolt split flange* (see Figure 4.32).

If the ports are internally threaded, they will probably contain either a pipe thread (NPTF for fluid power systems) or a straight thread with a seal. Until recently, the pipe thread was the most commonly used; however, pipe threads have several disadvantages—torque sensitivity, limited reusability, poor sealing at high temperatures, low vibration resistance, poor temperature cycling tolerance, and a tendency to gall (Perorazio 1988). These problems have led to the development of straight-threaded fittings using O-rings for sealing.

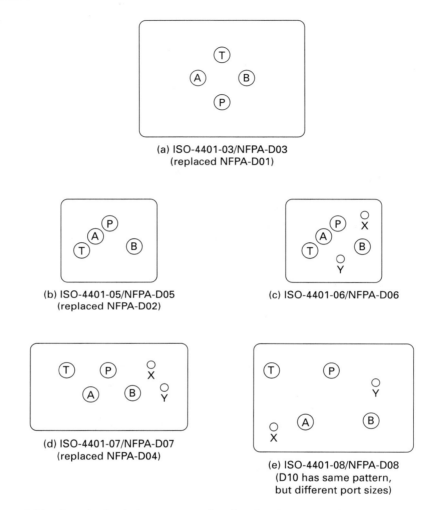

(a) ISO-4401-03/NFPA-D03
(replaced NFPA-D01)

(b) ISO-4401-05/NFPA-D05
(replaced NFPA-D02)

(c) ISO-4401-06/NFPA-D06

(d) ISO-4401-07/NFPA-D07
(replaced NFPA-D04)

(e) ISO-4401-08/NFPA-D08
(D10 has same pattern,
but different port sizes)

Figure 4.31   Standard subplate patterns for directional control valves.

The straight-threaded fittings come in two basic sealing configurations—the SAE O-ring boss and the O-ring face seal. The O-ring boss places the O-ring between the subplate surface (outside the thread) and either the tightening "hex" surface of the fitting or a locknut (on elbows, tees, etc.). The face seal fitting has an O-ring on the end of the fitting that mates with a sleeve on the conduit or with the bottom of the port hole. These O-ring seal designs are tolerant of surface imperfections, compensate for temperature fluctuations, are easy to install, and are virtually leak free. The locknuts on elbows and tees allow them to be positioned as desired without over- or undertorquing. There are numerous O-ring materials available to ensure fluid compatibility. Since there is no need for a thread sealant, the chance of fluid contamination is significantly reduced. These straight-

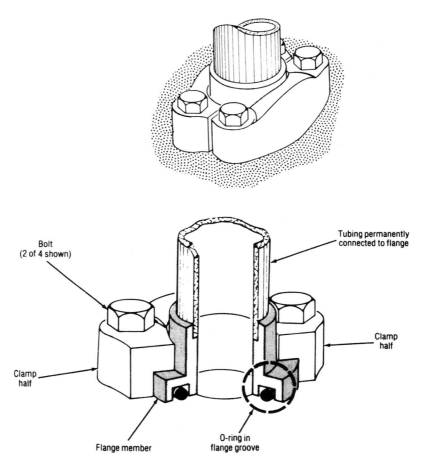

Figure 4.32 SAE split-flange O-ring port fittings are widely used on heavy equipment such as tractors and bulldozers. (Courtesy of Penton/IPC)

thread fittings with O-ring seals are a much better choice for most applications than the pipe thread.

The third connector is the four-bolt split flange shown in Figure 4.32. This type of fitting requires that the conduit be prepared by the installation of an attachment fitting (usually brazed or welded to the pipe or tubing). This fitting has a clamping ring and an O-ring seal groove. The two pieces of the split flange are bolted to the subplate surface over the clamping ring and torqued to retain the ring and compress the seal.

The split flange is especially suitable for large pipe sizes and high pressures. It requires more time and care in assembly and a clean sealing surface, but it is good for installations where space is limited, is easy to position, and has long service life (Perorazio 1988).

4.8   TRANSDUCERS FOR SOLENOID SYSTEMS

A transducer can be defined as any device that changes a mechanical parameter to an electrical signal. Because solenoid valves are simple off-on devices, they are usually used to change linear or rotary position, rotary speed, liquid level, pressure, and the like, into a signal that can be used to control the solenoid valve (either directly or through a programmable logic controller (PLC), as we will see later).

The transducers used as control elements are themselves simply switches that respond to a system parameter to either complete or break the control circuit. The majority of these switches sense a discrete position and are known as *limit switches*. Others, called *pressure switches*, are used to sense high or low pressure or vacuum limits. Centrifugal switches may be used to sense overspeed or underspeed in hydraulic motors. Level switches may be used to monitor reservoir or product feed levels and to open or close valves in response to the detected level. The operation of these switches and sensors will be discussed in more detail in Chapter 8.

4.9   TYPICAL SOLENOID CIRCUITS

In this section we will look at some very basic solenoid circuits. Although these circuits are somewhat simplistic, the concept of solenoid control circuits is sufficiently illustrated to allow you to undertake more difficult circuits with a degree of confidence and competence.

The most basic solenoid control circuit is one in which the operator uses a switch to alternately energize and deenergize the solenoid. A momentary push button or any number of other switch types may be used to accomplish this task. Let us begin with the momentary push button.

The relay logic diagram and the hydraulic circuit are shown in Figure 4.33. In this circuit we have a normally open push button and a four-way, two-position, single-solenoid, spring-returned valve. When the solenoid is not actuated, the spring pushes the spool into its unactuated position, causing the cylinder to retract. When the operator pushes the push button, the electrical circuit is completed, and solenoid A is actuated. This shifts the valve spool and causes the cylinder to extend. When the operator releases the push button, the circuit is broken. This deenergizes the solenoid and allows the spring to return the spool to its unactuated position. The cylinder retracts with the spool in this position. The cylinder will remain in its extended position only as long as the operator holds the button down. (Notice the cross-referencing between the ladder diagram and the hydraulic circuit diagram.)

We can express the logic of the circuit of Figure 4.33 in words by saying "If push-button A is pushed, then solenoid A will be energized." If we designate the action of pushing the button as simply "A" and the energizing of the solenoid as "B," we can reduce this statement to "If A, then B." We can further reduce this to a simple equation by using Boolean algebra.

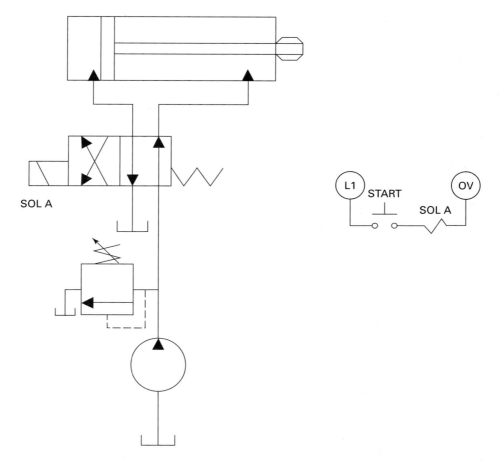

Figure 4.33 In this circuit the START pushbutton must be held down to hold the cylinder in the extended position.

Boolean algebra was developed by George Boole and presented in his paper "Mathematical Analysis of Logic" in 1847. Originally, Boolean algebra was an abstract mathematical system, but modern-day applications can be found in virtually any area where cause and effect are discrete functions. Digital electronics and fluidics are two such areas. Here, we are applying the technique to simple switching elements. Let us digress slightly and see how this algebra of logic works. Don't worry; we will just skim the surface.

The variables in Boolean algebra are designated by the letters *A, B, C,* etc. Because each letter represents an output event, we can assign a digital value to the event. We use a 1 to represent an output signal (electrical high) and 0 to represent the absence of an output signal (electrical low). We show the complement (or opposite) of a variable by placing a bar over the letter. For instance, if we designate A to be an output, then $\overline{A}$ indicates that there is no output. The digital equivalent would be A = 1, $\overline{A}$ = 0.

Figure 4.34 Series toggle switches to control a light (AND circuit).

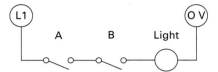

As you might expect, certain mathematical signs are used in Boolean equations. The meaning of the signs is, however, different from the normally accepted mathematical usage. These mathematical signs and their meanings are as follows:

a. A dot (·) indicates that the subsequent output is required *in addition* to the preceding output—in other words, a *series* arrangement. It is equivalent to an AND-gate logic element. In an equation a dot is read as the word *and*. The absence of any sign between variables implies the presence of the dot.

b. A plus (+) indicates an alternate logic; the subsequent output may be used *instead of* the preceding output—a *parallel* arrangement. It is equivalent to an OR-gate logic element. In an equation it is read as the word *or*.

c. The equals sign (=) indicates a resultant response. In an equation, it is read as the word *then*.

Let's look at a few simple equations and their literal meaning. The equation A·B = C (or AB = C) means, "If A occurs *and* B occurs, then the result will be C." An equivalent statement is, "IF A AND B THEN C." This is actually an "IF AND ONLY IF" statement. Both A and B are required if C is to occur.

We can apply this equation to a switching circuit as shown in Figure 4.34. In this circuit we have two normally open switches: A and B, and a light, C. Both switches must be closed if the light is to come on.

If we change the circuit by making B a normally closed switch (Figure 4.35), then we will need to actuate only A (but not B) to turn on the light. We write the Boolean equation for this circuit as A $\overline{B}$ = C. "If A is actuated and B is not actuated, then C will occur." We would read this as "IF A AND NOT B THEN C."

You may have noticed by now that the AND logic is associated with series circuits. In such a circuit, every device must be positioned so as to complete the circuit to energize the output device (the light in our example circuit). This means that all initially open devices must be closed and all initially closed devices must remain closed.

A parallel circuit, in contrast, offers alternative current paths so that the completion of any one path will cause the output device to be energized. Figure 4.36 is an example of a parallel circuit in which the closing of either switch (A OR B) will turn on the light. The Boolean equation for this circuit is A + B = C which is read, "IF A OR B THEN C."

Figure 4.35 A series circuit as in Figure 4.34 except toggle switch B is normally closed.

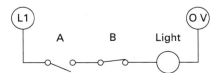

Figure 4.36 Toggle switches in parallel (OR circuit).

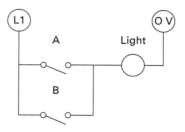

With the concept of Boolean algebra in mind, let's continue our discussion of circuits for energizing our solenoid. How can we modify the control circuit of Figure 4.33 so that the cylinder will remain extended without the need for the operator to hold the push button down? We could accomplish this by using an electrical holding circuit through a relay. Although this type of circuit might be considered overkill for this simple task, we will look at it here because we will need the concept later. The hydraulic portion of the circuit will remain the same as in Figure 4.33, so we will alter only the control circuit for the time being.

Figure 4.37 illustrates the basic idea of using a relay to control the solenoid. In this circuit the normally open push button completes the electrical circuit to the coil of the contact relay (1CR). When the relay is energized, every set of contacts associated with that relay changes from its normal configuration to its actuated configuration. Thus the normally open contacts (1CR on rung 2) close, causing solenoid A to be energized. Unfortunately, this circuit accomplishes nothing more than the simple push button of Figure 4.33, because as soon as the button is released, 1CR is deenergized, the 1CR contacts open, and the solenoid is deenergized. So it appears (correctly) that we have gained nothing except complexity. The Boolean algebra for the circuit logic will show this.

For the first line of logic we have

$$A = B$$

For the second line we can write

$$C = D$$

Because C represents the contacts of relay B and will respond to the energizing of B, we can also write

$$B = C$$

Figure 4.37 The START button energizes the relay that energizes the solenoid.

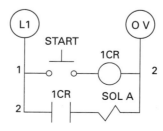

Thus

$$A = D$$

That is, if A is actuated, D will be energized. It also means that if A is not actuated, then D will not be energized, so we have gained nothing—we still have to hold the button down.

How can we fix this problem? Let's assume that there is another set of normally open contacts available in the relay. If we can tie this set of contacts into the circuit that energizes B (that is, 1CR), we may be able to use it to "hold" the relay. Figure 4.38 illustrates this concept.

In this circuit the contacts of 1CR on rung 2 are connected in parallel with the push button. Consequently, energizing the relay by pushing the start push button causes these contacts to close. This gives us a circuit that bypasses the push button, so even when the push button is released, the relay remains energized. Since this holds the 1CR contacts on rung 3 closed, the solenoid remains energized.

We can express the logic of this circuit as A + E = B, and C = D. But since B = C, we can also write A + E = D.

There is one minor problem with the circuit in Figure 4.38. What is it? How do we deenergize the solenoid to allow the cylinder to retract? We can't! We failed to provide that option.

The solution is simple. We need only to put in another switch in series between the parallel circuit and the relay to momentarily deenergize the relay. We use a normally closed push button in our circuit, although other types of switches could be used. Figure 4.39 shows this circuit, and we can see that as long as either the START push button or contacts 1CR on rung 2 are closed and the STOP push button is not actuated, the relay, and consequently the solenoid, will be energized. Pushing the STOP button breaks the circuit, deenergizing the relay and solenoid, and retracting the cylinder. In Boolean logic, we express the first line as (A + E)F = B. Does this circuit look familiar? It should. It is exactly the same as the circuit in Figure 3.20, except with a solenoid instead of a motor.

Now, let's look at a circuit in which we incorporate some automation to supplement the manual control illustrated above. In Figure 4.40 we see a system in which the operator initiates a cycle by pushing the START (momentary) push button. The solenoid is energized, allowing the cylinder to extend until the limit switch (1LS) is contacted. This

Figure 4.38 A holding circuit added to Figure 4.37.

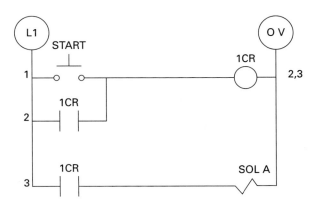

Figure 4.39 A relay holding circuit with a STOP push button.

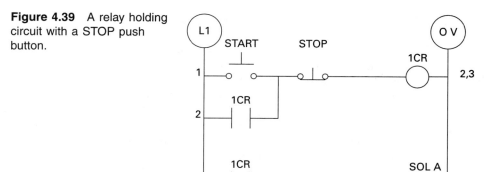

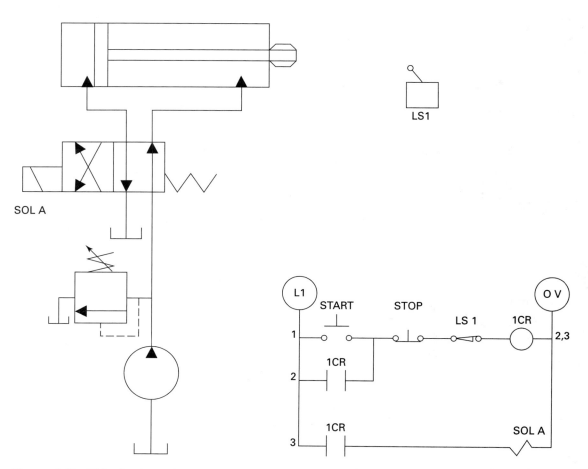

Figure 4.40 This circuit provides automatic, one-time reciprocation of the cylinder using a limit switch.

breaks the circuit to the relay and deenergizes the solenoid. The spring shifts the spool, which causes the cylinder to retract.

Notice the importance of the placement of the limit switch in the control circuit. If it was in series between the 1CR contacts on rung 3 and solenoid A, it would deenergize the solenoid when it opened; however, as soon as the cylinder began to retract, the limit switch would close again. This would reenergize the solenoid and cause the cylinder to extend once again, resulting in the cylinder's reciprocating at the end of its stroke. If the limit switch were located in series with the START push button but prior to the junction of the parallel circuit containing 1CR contacts on rung 2, it could not cause the relay to be deenergized. Notice, again, that the nomenclature used in the control and hydraulic circuit diagrams must coincide to facilitate a logical and convenient understanding of these two separate *circuits* as an electrohydraulic *system*.

Figure 4.41 illustrates a timer control circuit. Notice that the hydraulic circuit remains unchanged from the earlier illustrations. The difference in the operation is in the

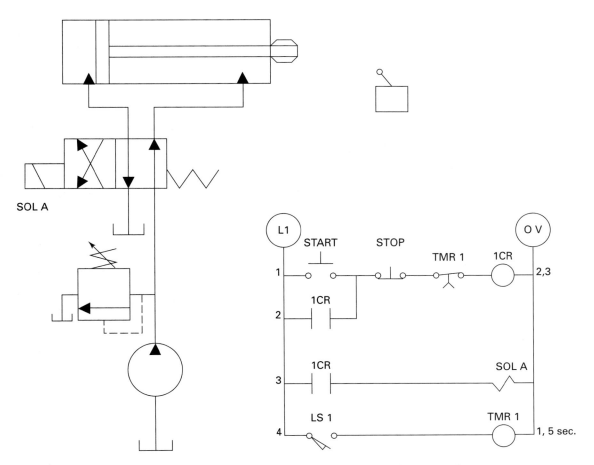

Figure 4.41 The addition of the timer causes the cylinder to delay before retracting.

control circuitry. Here the normally open limit switch is used to initiate the timer. When the cylinder extends and contacts 1LS, the timer (TMR 1) is activated. The cylinder remains fully extended until the timer runs out and the normally closed contacts associated with it (TMR 1 on rung 1) change to the open state. This action deenergizes the relay (1CR), which, in turn, opens all 1CR contacts, deenergizes the solenoid, and allows the cylinder to retract. When 1LS opens, the timer resets itself in preparation for the next cycle.

Obviously, these circuit concepts can be extended ad infinitum. You will be given the opportunity to use your skills and imagination to develop additional circuits by completing the practice exercises. Remember, there are usually several ways to accomplish most of the required tasks.

4.10 SOLENOID VALVE MAINTENANCE AND TROUBLESHOOTING

Solenoid valves are rather simple devices that require little maintenance. As far as the valve hardware (spool, body, etc.) is concerned, there are only two major considerations. The first, which really concerns design rather than maintenance, is to initially size the valve properly. An undersized valve leads to two problems: excessive pressure drop which leads to fluid heating and unnecessary energy costs; and high flow velocities through the valve, which can cause cavitation and erosion damage to the valve lands.

The second consideration is contamination control. There are four main contaminants to be aware of—solid particles, chemicals (including water), air, and heat.

Solid particles can cause the valve spool lands and the valve bore to wear. This leads to higher internal leakage between the valve ports and detracts from valve performance. Solid particles can also cause the valve to jam or hang open or closed. Most manufacturers recommend either a filter rating or a fluid cleanliness level to provide adequate protection for the valve. You should consider this a *minimum* requirement and provide cleaner fluid if possible (and it usually is possible).

Chemicals can cause the valve to rust, erode, or corrode, which can result in either excessive internal leakage or jamming.

Air in the hydraulic fluid can lead to two problems. The first is cavitation damage, which blasts microscopic particles from the spool lands, bore, and port edges. The second is oxidation of the fluid (especially when the fluid temperature is excessive). Oxidation causes the formation of sludges, gums, and varnishes, which can cause the spool to stick in the bore.

Heat causes the fluid to deteriorate in many ways, oxidation probably being the most critical for solenoid valves. Others include temporary and permanent changes in viscosity, depletion of additives, acid formation, and changes in the chemical structure.

The solenoids that operate the valves are also rather simple. This simplicity ensures a long and reliable life if they are properly installed, used, and maintained. In fact, most solenoid manufacturers advertise the lives of their products in millions of cycles, which represents years of use for many industrial applications.

AC solenoids tend to be more susceptible to damage and failure than DC solenoids, due to the high inrush current that we discussed earlier. The high current generates high heat, which causes insulation failure.

Problems that cause the inrush current to be maintained include:

- *Failure of the plunger to close.*
- *Valve spool sticking.* This is usually the result of solid contaminants or sticky residues between the spool and the bore.
- *The simultaneous activation of both solenoids on double-solenoid valves.* The control circuitry should be designed to prevent this. The individual circuits can also be provided with fuses to protect the solenoids.
- *Low applied voltage.* If the voltage is too low, the solenoid cannot generate sufficient force to move the spool, although there may be a high rate of heating.
- *High cycling rate.* The heat generated by the repeated inrush current may not be dissipated quickly enough to prevent overheating of the solenoid.

Other causes of AC solenoid failure include:

- *Overvoltage.* If the voltage is too high, the solenoid will close, but the holding current may still be excessive and result in coil burnout. This is an application problem.
- *High transient voltage.* This usually occurs when a common 120 V line is used for both the solenoid and a high inductive load such as an electric motor. Large voltage spikes can occur on motor disconnect.
- *Excessive solenoid force.* This can be the result of either overvoltage or oversizing of the solenoid. Failure occurs because the solenoid literally beats itself to pieces and fails structurally or "mushrooms" the pushpin, causing it to shorten enough to prevent complete stroking of the spool.
- *Corrosion in the solenoid.* This is a common problem in applications in a very wet or corrosive environment.
- *Ambient temperature too high.* Recall from our discussion of insulation temperature limits that ambient, fluid, and operating temperatures all affect the insulation.
- *Ambient temperature too low.* This can cause binding in structural components in the valve and solenoid as well as high resistance to spool movement due to high fluid viscosity.

Because the current through a DC solenoid is the same regardless of the plunger position, the inrush current problems do not exist. High voltages, excessive force, and high temperatures are the major causes of DC solenoid failures.

When the solenoid valve has a manual override pin, the separation of mechanical and electrical problems becomes very straightforward. If the valve is not working properly, simply push the pin. If the desired result is obtained, the problem is probably electrical. If the desired result is not obtained, the problem lies either in the valve hardware or elsewhere in the fluid power circuit.

It is poor maintenance practice to simply replace a valve or a solenoid without determining the cause of the failure. To do so may lead to another failure in an unacceptably short time. Table 4.2 lists the symptoms and possible causes for solenoid failures.

TABLE 4.2 Causes of and Cures for Solenoid Burnout

Observable Damage	Possible Causes	What Happens	Remedy
Coil insulation is burned; plunger is in open position; nylon coil bobbin is melted under the plunger	1. Low line voltage	Insufficient force to close plunger; high inrush current continues and generates excessive heat[a]	Replace coil only[b]; correct low voltage
	2. Ambient temperature too high	Plunger eventually will not close because undissipated heat reduces current flow (closing force) while electrical resistance increases and generates more heat[a]	Replace coil only[b]
	3. Cycling too fast		Install a continuous-duty model
	4. Load too high, or valve spool is blocked	Plunger blocked in open position permits electrical resistance to increase and generate excessive heat[a] (see also note below)	Replace coil only[b]; correct high-load condition
Coil insulation is burned; plunger is in closed position (no melted nylon)	1. Voltage too high	Extra force of excessive holding current holds plunger in closed position while electrical resistance increases, generates excessive heat, and burns out coil	Replace coil only[b]; correct high-voltage condition
	2. Solenoid too large for light load		
Frayed and burned lead wires	External mechanical short	Water-based coolant, metal chips, etc., have created contact between wires	Replace coil only[b]; shield from coolant
Small pinhole burn in coil wrap	Transient short to ground	High-voltage surge causes spark to jump between coil winding and solenoid C-stack (or other nearby ground)	Replace coil only[b]
Spongy insulation on coils and lead wires	Internal mechanical short	Fire-resistant fluids (phosphate esters) dissolve coil insulation, coil varnish, paint, etc.; cause short between coil turns	Install solenoid with proper insulation
Deep scoring at all seating surfaces	Overvoltage or reduced load (or wrong size)	Excessive closing force causes T-bar to wear through copper shading coils at top of C-stack; plunger also hammers base, resulting in destruction	Replace the entire solenoid, not just the coil

[a]Excessive heat burns insulation off coil wires, permitting electrical short and melting nylon bobbin.
[b]If solenoid is the encapsulated type, the entire solenoid must be replaced.
Note: In the case of double-solenoid valves, the cause of solenoid burnout may be that both solenoids were actuated simultaneously. Usually, one solenoid will burn out because it is unable to close properly.

Each failed solenoid should be examined, using this listing to determine the exact causes for the problem so that appropriate action can be taken to prevent failures.

4.11   SUMMARY

Solenoid valves are the most commonly used electrohydraulic and electropneumatic valves. They are relatively simple devices consisting of a solenoid (usually of the wet-armature type) that is used to position the valve spool by pushing against a pushpin. Solenoid valves may be direct-acting (in which the solenoid plunger moves the valve spool directly) or pilot operated (in which the solenoid plunger moves a large main spool indirectly by providing "power steering" through a small pilot valve).

The speed with which direct-acting solenoid valves shift is a function of several parameters including initial solenoid force, spool mass, and the solenoid force gradient. They will generally shift fully in 16 to 22 ms. Pilot-operated valves shift much more slowly (150 to 200 ms) because of the large spool mass, viscous drag, spring forces, and other inhibiting forces.

The electrical control circuitry we discussed in Chapter 3 can be extended to provide control for solenoid valves. In this chapter we expanded the control circuit concepts to include timers and limit switches. You will see even more variations as you draw the control circuitry required for the review problems.

The solenoid valves discussed in this chapter were "bang-bang" valves, meaning that they are either open or closed with no intermediate stroke capability. In the next chapter we will examine an advancement in solenoid technology that provides a great deal more flexibility.

REVIEW PROBLEMS

General

1. List the three basic components of a solenoid.
2. What is the purpose of a shading coil?
3. Explain the conceptual difference between an air-gap solenoid and a wet-armature solenoid.
4. Why are wet-armature solenoids preferred for hydraulic valve applications?
5. What is the difference between the inrush current and the holding current of a solenoid?
6. Under what circumstances is the inrush current potentially harmful to a solenoid?
7. Is inrush current important in DC solenoid technology? Why?
8. List and describe briefly the standard ISO subplates used for directional control valves.
9. Why is a float-center pilot valve normally used with a spring-centered main spool?
10. Why are pilot-operated main spools used in solenoid valve circuits?

Electrical Calculations

11. A solenoid valve uses a plunger with a diameter of 0.5 cm and a stroke of 0.7 cm. The 2500-turn coil has an inrush current of 2.0 A. Find the force available to move the spool.

12. A 120 VAC solenoid has a holding current of 0.6 A. What is the power (in watts) of the coil?

13. A coil produces an initial force of 1 lb and a final force of 8.5 lb. It has a stroke of 0.375 in. Find the force gradient of the solenoid and its force after it has moved 0.25 in.

14. A solenoid valve uses the coil of Problem 13. The spool has a mass of 0.0002 lb·s^2/in. How long does it take the spool to shift through its full stroke?

15. A pilot-operated valve uses the valve of Problems 13 and 14 to operate the main spool. The pilot pressure in the system is 200 psi. The total back pressure is 65 psi. The pilot spool has an effective orifice area of 0.055 in.^{2} with a flow coefficient of 0.68. The main spool is 2 in. in diameter and has a 1.8-in. stroke. The fluid density is 8.5 × 10^{-6} lb·s^2/in.^{4}. Find the total time required to shift the valve.

Circuit Practice

In the following problems, assume the use of standard eight-pin relays. Use standard graphic symbols to draw both the electrical ladder diagrams and the hydraulic circuits. Provide full documentation for the ladder diagram and cross-referencing between the ladder diagrams and the hydraulic circuits.

16. A double-acting cylinder extends to a preset point, then retracts automatically.

17. A bidirectional motor rotates in one direction for 30 s, then reverses for 30 s, then stops.

18. Two double-acting cylinders are required to operate in sequence. Draw the circuits required for the following sequences:
    a. Extend 1, extend 2, retract both
    b. Extend 1, extend 2, retract 2, retract 1
    c. Extend 1, extend 2, retract 1, retract 2

19. A system is required to operate at 500 psi during the cylinder stroke, then at 1000 psi while the cylinder is at full stroke. The cylinder remains at full stroke for 20 s, then retracts automatically. The pressure goes back to 500 psi for retraction.

20. A small pneumatic cylinder is used to check the orientation of rectangular parts on a conveyor. If the part is improperly oriented, its short dimension will be vertical. In this case, it must be rejected (pushed off the belt). Draw a system to do this job.

REFERENCES

1. Detroit Coil Co. Solenoid Application Data; AC Solenoids on DC. Ferndale, Mich.
2. Detroit Coil Co. Solenoid Application Data; Transient Suppression. Ferndale, Mich.

3. Detroit Coil Co. Solenoid Application Data; 50- and 60-Cycle Solenoids. Ferndale, Mich.

4. Detroit Coil Co. What Is a Solenoid?, Bulletin 191. Ferndale, Mich.

5. Parker Hannifin Corp. Bulletin 7325, Choosing the Right Valve. Cleveland, Ohio

6. Perorazio, David. 1988. How to Prevent Hydraulic System Leakage. *Plant Engineering*, December 15.

7. Whitmore, Chuck. "The Wet Armature Solenoid: Birth of a New Generation." 1975–1976 Fluid Power Technology Conference, Chicago, Ill.

8. Yeaple, Frank. 1990. *Fluid Power Design Handbook*. New York: Dekker.

Suggested Additional Reading

Johnson, Jack L. 1992. *Basic Electronics for Motion Control*. Cleveland, Ohio: Penton.

CHAPTER 5

Electrohydraulic Proportional Control Valves

OBJECTIVES

When you have completed this chapter, you will be able to:

- Explain the differences between proportional control valves and servovalves.
- Explain the operation of proportional solenoids.
- Discuss the advantages of proportional control valves over standard solenoid valves.
- Recognize and draw the ISO symbols for proportional control valves.
- Explain the use of proportional control valves in directional, flow, and pressure control applications.
- Discuss the differences between force control and position control.
- Explain the concepts of the electronics required to operate proportional control valves.
- Explain pulse width modulation.

5.1   INTRODUCTION

Electrohydraulic proportional control valves (EHPV) offer a step up in the technology of fluid power controls. They move us from the relatively low-tech world of simple on-off controls into a higher technological plane in which a more sophisticated valve is operated by electronics rather than just electrical switching. The advantage of this step is greater flexibility in system design and operation as well as a decrease in fluid-power circuit complexity for processes requiring multiple speed or force outputs.

Because of the extreme flexibility, the combinations of functions (such as flow control and directional control in a single valve), and the rapid cycling capabilities of modern EHPVs, the applications for such valves are continuously expanding. An example of such an application is in controlling the acceleration and deceleration of a conveyor belt.

Conveyor belts are commonly driven by hydraulic motors and are used, for example, in paper mills where commercial bond paper is produced. An overall diagram of the system is shown in Figure 5.1. In the last stages of production, the paper is stacked on a conveyor belt that moves the stacks into the area where the paper is wrapped for shipping. The movement is indexed so that a stack is moved into place, and the belt stops while the bundle is wrapped and then replaced on the belt to be carried to the loading area. The belt speed must be maximized, but due to inertial effects, sudden starts and stops must be avoided so that the stacks of paper do not become skewed or topple completely. A flow control EHPV is used for this purpose, with a speed ramping function providing the controlled acceleration and deceleration.

A photoelectric sensor provides the control switching to start and stop the valve. As the paper stack breaks the beam a stop command is issued, and the ramp closes the valve at a predetermined rate. When the stack is removed for wrapping, the beam is remade, but the control algorithm is such that the start command does not occur until the paper bundle is replaced and the beam is broken again. This signals the valve to ramp open, causing the motor to run at the predetermined speed until the beam is broken by the next stack.

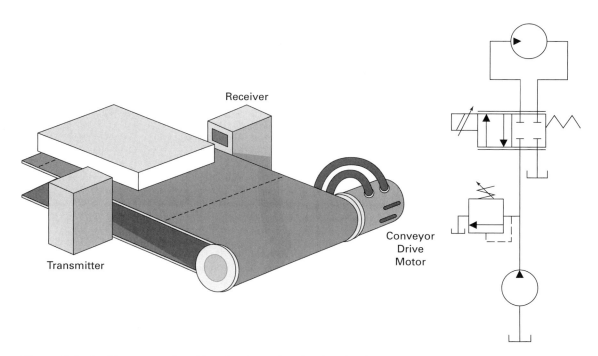

Figure 5.1 In this system, breaking the beam breaks the circuit to a proportional control valve. The motor slows to a stop under the control of the valve ramping function.

5.2   DEFINING PROPORTIONAL CONTROL VALVES

What is an electrohydraulic proportional control valve (EHPV)? This question is asked wherever electrohydraulics is discussed. The very broad definition states that "a proportional valve is any valve that produces an output (pressure, flow, direction) that is proportional to an electronic control input." Although this is a logical and correct definition, it is too broad for our purposes because it includes not only the valves we will discuss in this chapter but also the entire family of servovalves that we will discuss in the next chapter.

To help narrow the definition, and to illustrate some of the difficulty in attempting to define the valves we will be discussing, let's go back a few years to when these valves did not exist. At that time there were only two general types of electrically operated valves—solenoid valves and servovalves, and there was a huge performance gap between these two. The solenoid valve (as we have already discussed) was either actuated or unactuated—fully open or fully closed, with no intermediate position. Thus, it provided little of what we could actually term "control." It was simply an on-off valve, and its maximum frequency capability was usually 5 Hz or less.

The servovalve, in contrast, was a continuously controlled, high-frequency response device that received commands through its electronic control system that provided a high degree of control over position, velocity, acceleration, motor rotational speed, pressure, and other factors. It could accept and accurately respond to command signals at frequencies exceeding 100 Hz. The continuous feedback from the electronic transducers ensured high accuracy. Between these extremes, there was nothing—just a huge gap in performance, control capability, and cost.

That situation changed with the advent of the proportional control valve, often referred to as the electrohydraulic proportional valve. The design of its actuating device (the proportional solenoid, which we discuss later) allowed the spool to be stopped at intermediate positions rather than only at the ends of the solenoid stroke. The associated electronics controlled the spool position and offered a high degree of flexibility compared with the operation of the solenoid valve.

This new valve had a maximum frequency response of around 10 Hz—fast compared with the frequency capability of the solenoid valve, but still very slow compared with that of the servovalve. There was no feedback from the circuit, so the controllability and control accuracy were poor compared with those parameters of servovalves but greatly exceeded anything that could be achieved with solenoid valves. The final result was a valve that stood comfortably between the solenoid valve and the servovalve in performance, cost, and complexity. It was easily recognized and defined.

That situation has changed as the performance and application of proportional valves have evolved. First, a spool position feedback loop was added. Next came improvements in the designs of the spools and the electronics, then came external feedback systems, higher frequency response, better performance in accuracy, hysteresis, deadband, threshold, and other parameters. In short, the proportional valves began to look more and more like servovalves in capability. Naturally, these performance and control improvements carried with them an increase in cost. They also blurred the distinction between servovalves and proportional valves.

As a result, performance and control are no longer distinguishing criteria. Rather, physical features such as design and manufacturing processes are the defining characteristics. For instance, proportional valves are operated by proportional solenoids (to be discussed in the next section), whereas servovalves are operated by torque motors (to be discussed in the next chapter). The spools in proportional valves are almost entirely machine produced, while the spools for servovalves require a great deal of manual lapping and finishing. The clearances and tolerances in servovalves are much tighter than in proportional valves.

These differences in manufacturing requirements mean that the servovalve is still much more expensive than a comparably sized proportional valve. They also imply that

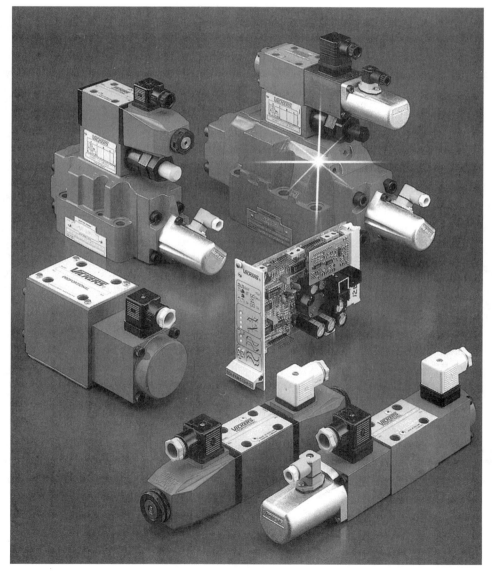

Figure 5.2 Proportional directional and throttle valves. (Courtesy of Vickers, Inc.)

the servovalve will, in general, outperform the proportional valve, especially in terms of accuracy, hysteresis, internal leakage, and other factors.

It is probably fair to say that a proportional valve can be likened to a low-cost, low-performance-range servovalve. To simplify our discussion in this chapter we will limit our definition of a proportional valve to any valve that is operated by a proportional solenoid rather than a torque motor (which is used for servovalves). We will divide these valves into three categories—directional, pressure, and flow controls—and take a look at each valve type separately. We will also define what we mean by a proportional solenoid.

5.3 DIRECTIONAL CONTROL EHPVS

The directional control valve is the most common EHPV. We will discuss this type in considerable detail because of its popularity. The general aspects of its operation also apply to pressure and throttle valves.

At first glance, the directional EHPVs shown in Figure 5.2 look very much like the solenoid valves we discussed in Chapter 4. Both types have solenoids (one or two), and both have a valve body with a movable spool, ports, and other components. There are, however, significant differences in both the solenoid and the spool. Let's look at these differences, beginning with the solenoid.

5.3.1 Proportional Solenoids

In our discussion of standard solenoid valves we noted that they have no intermediate positions. They are always at one end of the solenoid stroke or the other. This is inherent in the functioning of all standard solenoids. The magnetic flux attempts to drive the plunger to its fully closed position when the coil is energized. Recall that the force developed by the solenoid is a function of the square of the solenoid current and an inverse function of the square of the air gap (Equation 4.1). The result is that the force increases as the air gap closes as well as when the current is increased. Figure 5.3 shows

Figure 5.3 Typical solenoid force versus stroke curve at constant current.

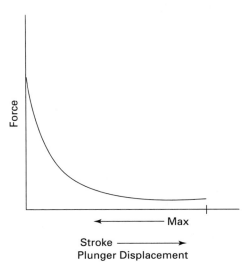

Figure 5.4 Solenoid force versus stroke curves with increasing current.

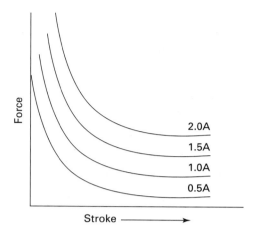

a typical force versus displacement curve. The solenoid force is at its minimum when the plunger is at its maximum position, and increases as the plunger moves in. By incrementally increasing the current in a particular solenoid, we can generate a family of curves as shown in Figure 5.4.

If there is a return or centering spring in the valve, the spring results in an additional design requirement for the solenoid. Let's assume we have a spring whose force is a linear function of its compressed distance ($F = kx$). If we plot the spring force versus the air-gap dimension (which is the same as the amount of spring compression), we get a graph like Figure 5.5. From this plot we can see that for the valve to operate at all, we must provide a current that will ensure sufficient solenoid force to overcome the spring force throughout the plunger stroke. If the solenoid force ever drops below the spring force, the solenoid will stop.

Figure 5.5 Solenoid force versus stroke curves with spring force overlaid.

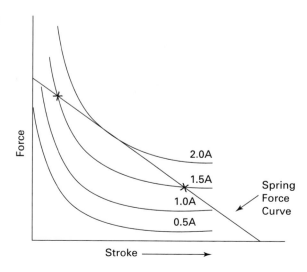

Figure 5.6 Proportional solenoid force versus stroke at constant current.

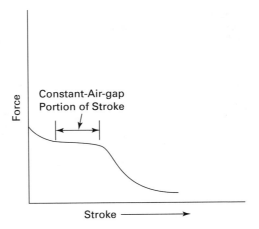

A proportional solenoid differs from a standard solenoid in the design of the area near the end of the plunger stroke. In the design of a basic solenoid, the air gap is closed at a uniform rate as the plunger moves in. Because the square of the air-gap dimension appears in the denominator of the force equation (Equation 4.1), the solenoid force increases exponentially as the air gap closes.

The design of a proportional solenoid eliminates the effect of the diminishing air-gap dimension at the end of the plunger stroke. This is accomplished either by utilizing construction features that actually maintain a constant effective air gap or by using magnetically impervious material that to the solenoid appears to be a constant air gap. The result for any given current is a force curve similar to that shown in Figure 5.6. The flat portion of the curve occurs in the constant-air-gap portion of the stroke. Varying the current results in a family of curves as shown in Figure 5.7. Use of a carefully designed, calibrated spring to oppose the solenoid force results in a solenoid force versus spring force arrangement like the one shown in Figure 5.8. The trick is to have the

Figure 5.7 Proportional solenoid versus force curve for varying current.

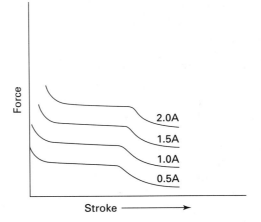

Figure 5.8 Proportional solenoid versus force curves with spring force overlaid.

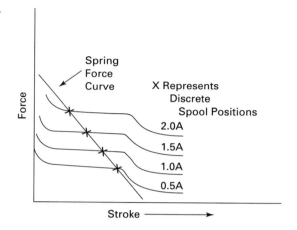

spring force curve intersect the solenoid force lines in the flat portion of the solenoid force lines.

The physical result of this arrangement is that the solenoid plunger position (and, subsequently, the valve spool position) can be controlled by the current applied to the solenoid. A higher current produces a higher solenoid force. This force compresses the calibrated spring until the spring force balances the solenoid force. When this force balance is achieved, the plunger stops.

The position at which the plunger (spool) stops determines the size of the flow path through the valve. This, along with the pressure differential across the valve, determines the fluid flow rate through the valve. The functional result is that both the flow direction and the flow rate can be controlled with a single valve. The portion of the solenoid stroke in which this occurs is usually referred to as the *control zone*. The length of this zone is only about 0.06 to 0.08 in. (0.15 to 0.20 cm). The total plunger stroke is also small, usually about 0.120 in. (0.3 cm).

5.3.2 Valve Hardware

Like any other directional control valve, EHVPs include a valve body, a spool, and miscellaneous piece parts as shown in Figure 5.9. The primary difference between the proportional and conventional valve hardware is in the spool design. The first difference (which is not obvious on visual inspection) is the high precision of manufacture of the spool. Straightness and roundness are carefully controlled (as close as 0.00005 in.). The radial clearance between the spool and the bore is usually three to four micrometers (0.00012 to 0.00016 in.).

The second difference is the metering notches on the spool lands. These notches, often referred to as *control grooves*, are used to provide a greater degree of control throughout the small spool stroke than would be possible with conventional square-edged spool lands. The number and shape of the control grooves play a major role in determining valve performance. Figure 5.10 shows a typical proportional spool.

As with conventional directional control valves, there is a large variety of proportional control valve configurations. These include both two and three positions, with open, closed, regenerative, and float centers in the three-position valves.

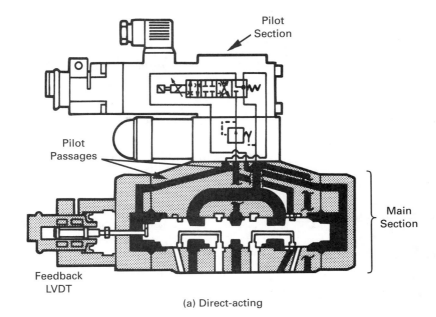

(a) Direct-acting

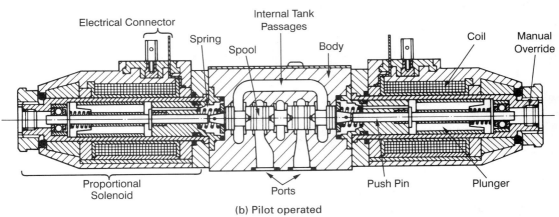

(b) Pilot operated

Figure 5.9   Typical EHPVs. (Courtesy of Vickers, Inc.)

The metering feature of the spools offers some additional functions not available in conventional solenoid valves. For instance, if all lands have identical control grooves, then flow is metered evenly in all directions. This is the most common configuration. By omitting some of or all the grooves on one of the lands, differential flows can be

Figure 5.10   Typical proportional valve spool configuration.

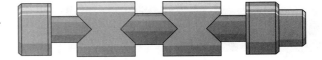

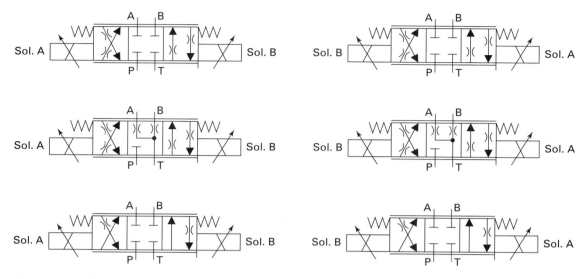

Figure 5.11 ISO symbols for a variety of proportional directional control valves.

obtained in the two directions. This concept can be used to cause a single-ended hydraulic cylinder to retract and extend at the same speed by reducing the input flow to the rod end on retraction. It can also be used to provide either meter-in or meter-out flow control.

Figure 5.11 shows the ISO symbols for several spool configurations. Notice the three additions to these symbols that are not seen in conventional solenoid symbols—the parallel lines on each side of the symbol indicate infinite positioning, the orifice symbols on the flow arrows indicate flow metering, and the diagonal arrows on the solenoid symbols indicate adjustment capability.

Proportional DCVs are all subplate mounted. The direct-acting valves use the ISO 4401-03 and -05 configurations. If larger valves are required to handle higher flow rate demands, pilot-operated valves are used. In proportional valve terminology, the pilot-operated valves are usually referred to as *two-stage valves*. Both the pilot and main stages of these valves contain metering spools.

5.4   THROTTLE (FLOW CONTROL) EHPV

Some EHPVs are designed to provide only flow control and not directional control. These valves effectively function in exactly the same way as a manually operated flow control valve. Figure 5.12 shows a cross section of a throttle valve along with its graphic symbols.

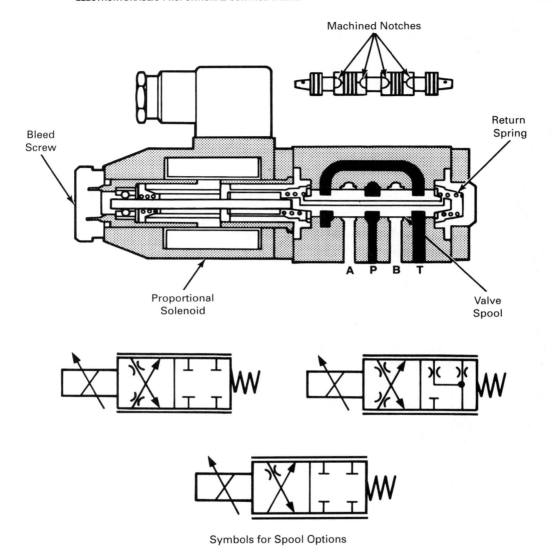

Figure 5.12 Cutaway drawing of a proportional throttle valve. (Courtesy of Vickers, Inc.)

As with the directional control valves, spool position is determined by the current input to the solenoid. Thus, the valves throttle flow in proportion to the command signal. As with any other such valve, the flow through the valve is also dependent on the pressure drop across the valve. The valve shown in Figure 5.12 has, by design, an inherent degree of pressure compensation. Thus, increasing pressure drop across the valve has progressively less effect on flow rate. Figure 5.13 illustrates this inherent compensation.

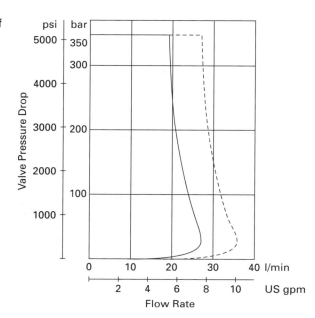

Figure 5.13 An example of pressure compensation in a proportional control valve.

5.5   PRESSURE CONTROL EHPV

Proportional-solenoid-operated pressure control valves can provide the same basic functions as conventional adjustable spring-type pressure control valves. These include pressure relief, pressure reducing, sequencing, unloading, counterbalance, and brake valve functions. The operation of the proportional valve is analogous to that of the standard spring-operated valve. The primary difference between the two is that current directed to the proportional solenoid determines the force exerted on the valve mechanism to hold the poppet (also referred to as the *cone* or *dart*) on its seat to prevent flow through the valve. This can be done in two ways. Some manufacturers omit the valve spring altogether and couple the solenoid plunger directly to the poppet through the push pin. Others retain the spring concept and use the solenoid to compress the spring to achieve the desired force.

Figure 5.14 shows an example of the latter type. In this valve the plunger of the proportional solenoid compresses the spring, which, in turn, seats the valve cone. System pressure is felt on the apex of the cone through the internal porting of the valve seat. As long as the solenoid-induced force on the back of the cone is greater than the pressure-induced force on the apex of the cone, the cone remains seated, and there is no flow through the valve. When the force holding the cone in its seat is exceeded, the valve opens and allows flow from port P to port T and back to the tank. The linear variable differential transformer (LVDT) shown on this valve is used to feed back valve position. This concept will be discussed later.

This particular valve design is termed *direct-acting* because the system pressure operates directly on the poppet. The flow-handling capability of a direct acting propor-

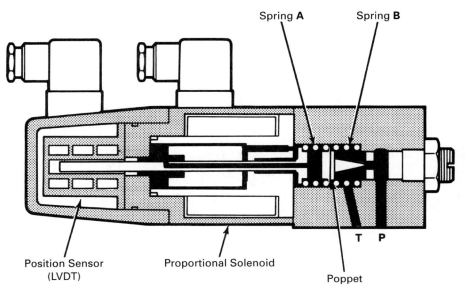

Figure 5.14 Poppet-type pressure relief valve with LVDT feedback. (Courtesy of Vickers, Inc.)

tional relief valve is severely limited (normally less than half a gallon per minute) because of the limited force capability of the solenoid. Proportional solenoids typically produce a maximum force of 15 lb or less, which represents the maximum force available to oppose the pressure-induced force on the poppet. The pressure-induced force is the result of the pressure multiplied by the area over which it is exerted ($F = p \times A$). Thus, only very small poppet areas can be used for the high operating pressures of modern fluid-power systems. It is this limited size that restricts the maximum flow through the valve.

If flows greater than the capability of the direct-acting valve are required, a pilot-operated valve must be used. In this unit the direct-acting valve discussed previously is used as the pilot section for the main spool.

Proportional solenoid relief valves provide significant advantages over standard relief valves because of their flexibility. They can provide a controlled rate of increase or decrease of system pressure to facilitate process demands. The use of the ramping function (to be discussed later) can ensure shock-free yet rapid pressure changes. The adjustable solenoid provides a remote control feature without the additional fluid piping and remote relief valve required for standard valves. The operations of the other types of pressure control valves are similar to the operations of the valves described here.

Figure 5.15 gives the symbols for several proportional pressure control valves. Notice that these are very similar to the symbols of their standard, nonelectronic counterparts but with the electronic control element replacing the manual or mechanical elements.

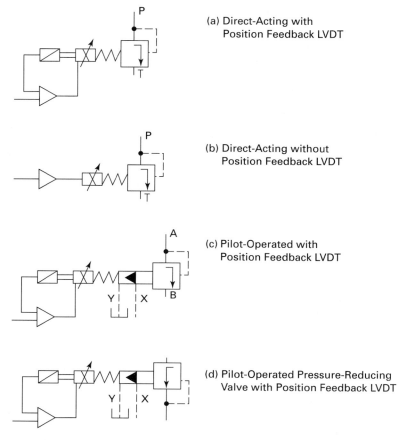

(a) Direct-Acting with
    Position Feedback LVDT

(b) Direct-Acting without
    Position Feedback LVDT

(c) Pilot-Operated with
    Position Feedback LVDT

(d) Pilot-Operated Pressure-Reducing
    Valve with Position Feedback LVDT

Figure 5.15   Symbols for pressure control EHPVs.

5.6   ELECTRONIC CONTROLS

Proportional control valves use modern electronic technology to provide the precise control of solenoid current. In this section we will look at the components and circuitry that provide this control. We are not going to try to make an electronics expert of you; that is left for other texts and educational experiences. Rather, we will consider the electronics to be gray boxes. (A gray box is similar to a black box, except that some light is shed on its functioning.) We will be far more concerned with what the electronic unit does than with how it does it. It is not necessary for you to understand the electronics in order to use them effectively; however, the more you understand, the more capable you will be of handling those situations that are out of the ordinary, especially in difficult troubleshooting and unusual design cases.

5.6.1 Force Control versus Position Control

In EHPV control terminology it is unusual to use the terms "open loop" and "closed loop," because of their traditional association with servovalves. In the broad definitions of these terms, however, both cases are actually applied to EHPV controls. (Recall that we used the open- and closed-loop terminology in Chapter 2 when we discussed fluid-power circuits.)

The term *open loop* when applied to electronic control circuitry implies that there is no feedback from the circuit to the control device. This is, in fact, the simplest control technique used for proportional valves. The operator—through the electronics—provides a command signal to the solenoid. The solenoid moves in response to that signal, repositioning the valve mechanism in the process. For example, as shown in Figure 5.16a the operator may use a potentiometer to send an input signal to the controller amplifier. The amplifier conditions the signal and sends the commanded current to the proportional solenoid. Ideally, this causes the solenoid to move to the commanded position, which, in turn, causes the output device to operate. If this output device is a hydraulic motor, it *should* turn in the specified direction and at the desired speed; however, in an open-loop system, there is no automatic and continuous monitoring of the motor to ensure that it is actually operating as intended. The operator is responsible for "eyeballing" the situation to observe direction and determine if the speed is acceptable.

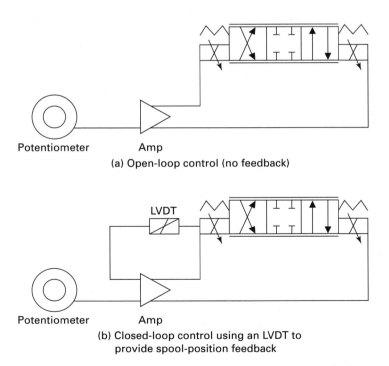

(a) Open-loop control (no feedback)

(b) Closed-loop control using an LVDT to
provide spool-position feedback

Figure 5.16 Proportional valves may be open-loop, or they may utilize an LVDT to provide a closed-loop system based on spool position.

Consider the case, however, where there are other opposing forces due to solid contaminants, varnishes or sticky residues as a result of fluid degradation, or any other situation that might impede the motion of the spool. These opposing forces have the same retarding effect on the spool as the calibrated spring. In these cases it is very likely that the *actual* final position will not be the *desired* final position, so the speed will not be the desired speed. It is then the responsibility of the operator to make the necessary control adjustments, because there is no automatic feedback mechanism to tell the electronics that the spool is not in the correct location.

This open-loop technology is usually referred to as *force controlled* in proportional valve terminology. The term implies that the position of the spool is determined strictly by resultant forces with no subsequent feedback or automatic adjustment to indicate the spool position.

In contrast, many proportional valves include a mechanism for automatically sensing the actual position of the valve spool. This actual position is then compared electronically with the commanded position. If there is a discrepancy, a supplemental signal is generated that repositions the mechanism as necessary. Note that this feedback is concerned with the position of the *valve* mechanism and has nothing at all to do with *load* parameters (position, speed, etc.). For this reason, EHPV circuitry employing this technique is usually referred to as *position control* or *stroke control* rather than closed loop.

In the closed-loop system of Figure 5.16b, the feedback loop does the monitoring automatically and continuously. Any needed corrections are made automatically. This requires more complex electronics plus the appropriate feedback mechanisms. Thus, it is more expensive but also more accurate and less operator dependent.

Recall from our earlier discussion of EHPV functioning that the position of the mechanism is that at which the solenoid force and the opposing force become balanced, giving a net zero force. In a directional control valve, for example, the calibrated spring provides the opposing force and therefore determines the final position of the spool. This, in turn, determines the direction and speed of the cylinder or hydraulic motor.

The feedback transducer that provides the information about the spool position is a linear variable differential transformer (LVDT). An LVDT, as shown in Figure 5.17, is a transformer with one primary winding and two secondary windings. The movable core is attached to the solenoid plunger and acts as a coupling between the primary and the secondaries. The position of the core determines the output from the secondaries. With the core exactly centered, the output is zero, because the secondaries are connected in series opposition so that equal but opposite outputs cancel each other. As the core moves, an imbalance in the outputs occurs, resulting in a net positive or negative output. This is the feedback signal from the internal closed loop that is fed into the control amplifier (which we discuss in detail a little later).

The amplifier electronics compare the feedback signal with the command signal to determine if the valve spool is in the commanded position. If it is not, an error signal is produced that is proportional to the positional error. This signal, in the form of an increase or decrease in the command signal that initially moved the spool, is used to move the spool toward the commanded position. This command-feedback-compare-adjust cycle is continuous, so that any deviation from the commanded position is immediately detected and corrected.

Figure 5.17 Linear variable differential transformer.

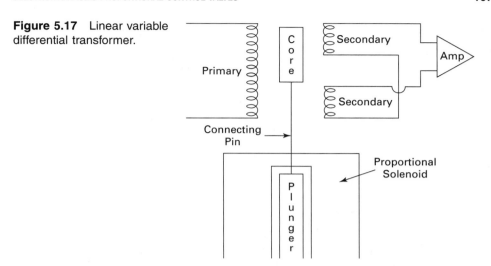

The valve shown in Figure 5.18 includes an LVDT as a position transducer. Typically, as shown in the figure, there is only one transducer, even though there may be two solenoids. In the case of pilot-operated directional control valves, there may be an LVDT on both the pilot spool and the main spool.

Obviously, the addition of the LVDT and the supporting feedback circuitry increases the cost of the system. The improvements in accuracy and overall valve performance and in overcoming the detrimental effects of fluid contamination, however, may well be worth the additional cost and complexity.

5.6.2 Electronics

The electronic units that control the force-controlled and position-controlled valves in their various configurations (directional, pressure, and flow control) are all very similar.

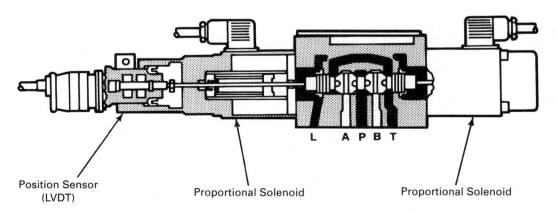

Figure 5.18 Proportional directional valve with LVDT feedback. (Courtesy of Vickers, Inc.)

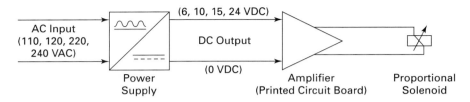

Figure 5.19 Block diagram of EHPV control electronics.

Basically, they consist of a power supply, an amplifier, and the proportional solenoid as shown in Figure 5.19.

Power Supply   All proportional solenoids operate on DC because it is somewhat easier to control than AC, and it prevents the heating associated with the operation of AC solenoids that are not allowed to complete their full stroke (refer to Chapter 4). The power supply is the DC power source for the system. Depending on the manufacturer, a power supply may utilize either a nominal 120 VAC or 240 VAC input. Using transformer stages it converts the input to one or more DC outputs ranging from 9 to 28 VDC, again depending on the manufacturer and the power requirements of the amplifier.

Power Amplifiers (Open Loop)   The power amplifier is the brains of the operation. It is a printed circuit board (PCB) containing the components necessary to provide and condition the various command signals to the proportional solenoid. Figure 5.20 is a photograph of a typical amplifier for controlling a directional control valve with two solenoids. A simplified circuit diagram of this amplifier is shown in Figure 5.21. The squares in this figure represent the various functions performed by the amplifier. In most cases a considerable number of the components shown in the photograph are required to accomplish the function represented by each functional symbol. Let's trace the power flow through the board and discuss the function represented by each of the symbols.

First, understand that everything contained within the solid rectangle of Figure 5.21 is physically located on the printed circuit board. The alphanumeric designators (b8, z10, d16, etc.) refer to pins on the PCB connector. Because the board is plugged into a card holder, in practice these designators specify terminals on the card holder terminal strips.

As a point of interest, the PCB connectors used for most proportional valves are designed along the lines of those commonly used in Europe; hence, these are referred to as *Eurocards*, and the holders as *Eurocard holders*. As a result, any Eurocard conforming to the German national standard designation (DIN) 41612 F28 will mate properly with any Eurocard holder conforming to that standard, regardless of the manufacturer of either component.

For the PCB in Figure 5.21, the 24VDC input is connected to terminals zbd 32 (+24V) and zbd 30 (0V). (The use of multiple alphabetic designators such as zbd indicates that the terminals are jumpered internally. Thus, in this case, the 24VDC connection could be made to terminal z32, b32, or d32 with the same result.) Additional signal conditioning and step-down transforming occur in the transformer (circle 1), and result in the availability of +10V, −10V, +15V, and −15V outputs at terminals z2, b2, z22, and b22, respectively.

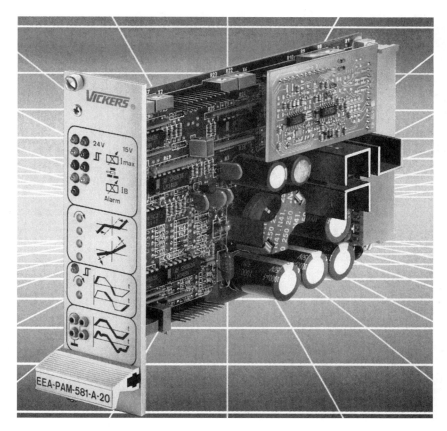

Figure 5.20 Typical amplifier printed circuit board. (Courtesy of Vickers, Inc.)

The outputs from terminals z2 and b2 are connected to one or more external potentiometers (circle 2) that are used to provide control inputs in the range ± 10V to terminals b6, b8, b10, and z8. Up to five external potentiometers can be used.

The significance of the positive or negative signal is that the signal polarity determines which solenoid will be energized. A positive voltage input will result in the activation of solenoid A, whereas a negative voltage will cause solenoid B to be energized.

The use of multiple potentiometers provides considerable flexibility and automation capability in the operation of the valve. For instance, if toggle switches are located in the potentiometer output lines, then the potentiometers can all be set to give different output voltages, both in magnitude and polarity. By alternately switching the potentiometers in and out of the circuit, both the direction and magnitude of spool movement can be controlled. The same result can be obtained with only one potentiometer by alternately changing its setting.

On reentering the board, the potentiometer inputs pass through the combinational amplifier (circle 3). This unit sums the inputs as necessary and provides polarity-sensitive gain. The gain can be adjusted using the gain potentiometers (circle 4) on the control panel

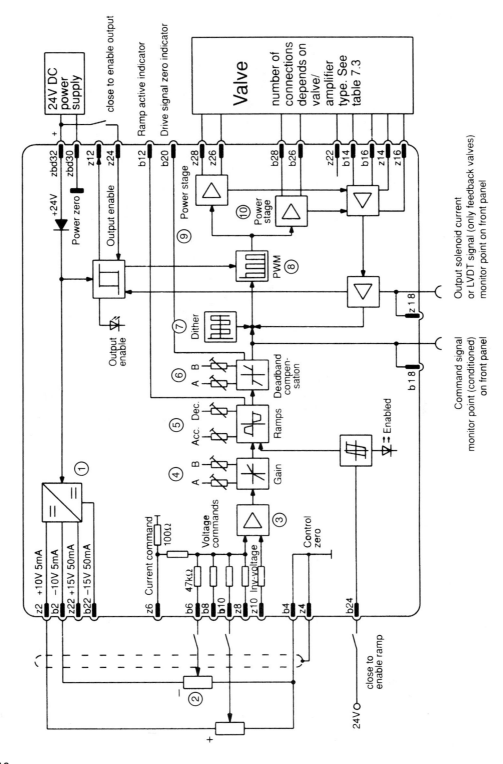

Figure 5.21 Block diagram and external connections for an EHPV amplifier. (Courtesy of Vickers, Inc.)

110

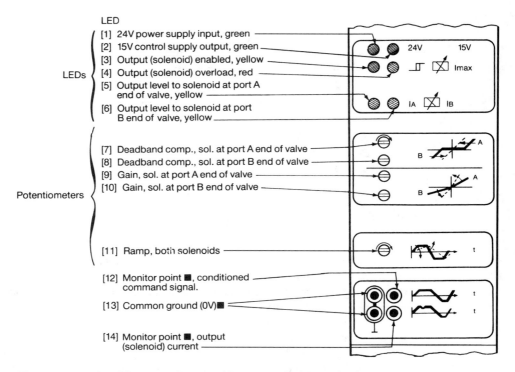

LED

[1] 24V power supply input, green

[2] 15V control supply output, green

[3] Output (solenoid) enabled, yellow

LEDs

[4] Output (solenoid) overload, red

[5] Output level to solenoid at port A
    end of valve, yellow

[6] Output level to solenoid at port
    B end of valve, yellow

[7] Deadband comp., sol. at port A end of valve

[8] Deadband comp., sol. at port B end of valve

[9] Gain, sol. at port A end of valve

Potentiometers

[10] Gain, sol. at port B end of valve

[11] Ramp, both solenoids

[12] Monitor point ■, conditioned
     command signal.

[13] Common ground (0V)■

[14] Monitor point ■, output
     (solenoid) current

Figure 5.22 Amplifier control panel. (Courtesy of Vickers, Inc.)

on the face of the PCB shown in Figure 5.22. The gains for solenoids A and B can be adjusted independently and can range from 30 mA/V to 320 mA/V.

The signal then passes to the ramp function controller (circle 5). This unit determines how rapidly the valve will move from one position to the next by controlling the rate at which the control signal changes from its initial value to its final value. This is accomplished by controlling the charge (or discharge) rate of a capacitor, which results in a corresponding increase (or decrease) in the voltage. For the maximum change of input signal (0 to $+10V$ or 0 to $-10V$) the time required for the valve to move through its full stroke can be adjusted for a minimum of 50 ms to a maximum of 2 s. For a smaller range of change, the time would be proportionately less for any ramp adjustment. Because only one ramp-adjust potentiometer is provided on the control panel, any setting controls both the ramp-up and ramp-down function for both solenoids. Some designs of controllers allow ramp up and ramp down to be controlled separately. Some designs also allow the ramping time to vary from nearly instantaneous to 5 s for full stroke.

Figure 5.23 illustrates the ramp function for four different input voltages. Notice that the absolute value of the ramp slope is always the same but that the duration of the ramp depends on the magnitude of the voltage change. For convenience (if not by design), we can refer to the positive voltage functions as "forward" speeds and negative voltage functions as "reverse" speeds. Thus, the operation shown in Figure 5.23 can be said to consist of two forward speeds and two reverse speeds. (Actual direction, of course, will depend on the hydraulic line connections.) Any combination of four forward and reverse

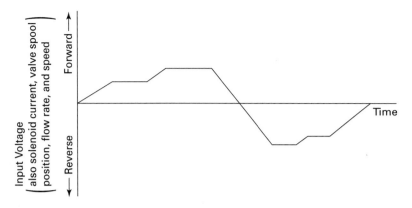

Figure 5.23   Valve operation with two forward and two reverse speeds.

speeds can be used. In fact, some clever crossover switching of the outputs from terminals z2 and b2 could actually provide eight possibilities. If a step function rather than a ramped change is desired, the ramp function can be disabled, which will reduce the change time to approximately 1 ms.

The next function provides deadband compensation (circle 6) by means of a function generator. The flow versus current characteristics of a typical valve are shown in Figure 5.24. The plot shows that there is no flow through the valve until the input current to the solenoid reaches about 25% of the maximum signal. The region of no flow is termed the *deadband* of the valve. It is a combination of several factors—both electronic and mechanical—including valve overlap, electrical threshold, hysteresis, friction, and other fac-

Figure 5.24   A typical proportional control valve flow output versus current input signal curve.

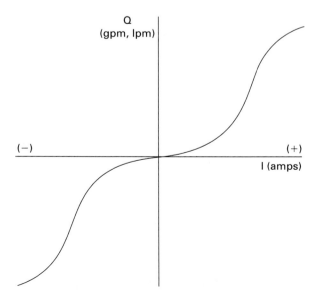

tors that delay or impede the initial motion of the valve. Although a certain amount of deadband can be desirable in proportional valves, the actual magnitude can be reduced (or increased, if desired) by adjusting the deadband compensation potentiometers on the control panel. The adjustment range is from 0 to 50% of the maximum signal. The reduced deadband is achieved by jumping the current to the preset value when the input signal goes either slightly positive or negative compared with the 0 V reference.

As the command signal passes the deadband compensator a 100 Hz dither signal is superimposed on it. This signal is of very low amplitude. Its purpose is to keep the solenoid in low-amplitude motion to reduce static friction and improve its response to signal changes. The dither signal generator (circle 7) also contains a triangle signal generator that is used as an input to the pulse width modulator (circle 8).

The modulator uses a technique known as *pulse width modulation* (PWM) to produce the current going to the power stage amplifiers. The concept of PWM is illustrated by Figure 5.25, which shows a stepped current pulse of a fixed amplitude generated at a

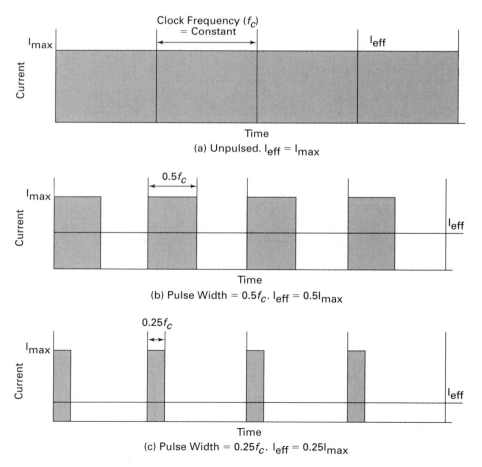

Figure 5.25 Pulse width modulation.

fixed interval. If the pulse is maintained throughout the interval (that is, until the next pulse is generated), then the average current and the maximum current are the same. If, however, the pulse is terminated before time for the next pulse, then the average current is reduced proportionately. The duration of the pulse within the frequency interval is termed the *pulse width*. The termination of one pulse and the regeneration of the next pulse is called *modulation*—thus the name pulse width modulation. As the duration (width) of the pulse decreases, the average (or effective) current output decreases. This effective current is then fed to the power stage amplifiers.

The signal then goes into the power stage amplifiers (circles 9 and 10), where components direct the command signal to either solenoid A (positive input) or solenoid B (negative input). If the input signal is negative, a blocking diode in power stage B (9) rejects the input, and no signal is passed to solenoid B. An inverter in power stage A (10), however, inverts the signal (changes it to positive) so that the blocking diode in that stage will pass it. The stage then processes the signal further and outputs it to solenoid A, which is connected to terminals z26 and z28. A positive input signal is inverted in power stage A so that it becomes negative and is rejected by the blocking diode; however, it passes through the diode of power stage B and subsequently powers solenoid B connected to terminals b26 and b28.

Light-emitting diodes (LEDs) on the control panel in Figure 5.22 provide status information. The actual command and solenoid signals can be monitored using the meter sockets provided on the control panel.

If a force-controlled valve, such as the one shown in Figure 5.9a is used, only terminals b26, b28, z26, and z28 of the PCB in Figure 5.21 are used for connecting the valve. However, for stroke-controlled valves, the LVDT must also be connected. Terminals z22, b14, and b16 are used for the LVDT on single-stage valves such as the one shown in Figure 5.18. A two-stage valve will require that terminals z14 and z16 be used also.

This amplifier can be used to control single-solenoid as well as double-solenoid valves. Terminal connection instructions must be carefully observed in all cases. The amplifiers for flow control valves and pressure control valves are similar to these.

As with standard solenoid valves, external signal switching is normally digital (discrete on-off) and is accomplished through the use of limit switches, pressure switches, toggle switches, optical or proximity sensors, or similar controls. Programmable controllers can be used to provide additional flexibility.

5.7   COMBINING SPEED AND DIRECTIONAL CONTROL

One of the major advantages of the EHPV is its ability to provide flow (hence speed) control as well as directional control in addition to controlling acceleration and deceleration. Although it is possible to achieve speed and directional control with a series of solenoid valves and flow control valves, a single EHPV can replace numerous conventional valves and combine all the control functions into a single unit.

In some surface-finishing operations, it is desirable to use different wheel speeds, starting with a low speed with a coarse-grit slurry and progressing to a high speed with a fine-grit slurry. This can be accomplished by taking advantage of the multispeed capabilities of an EHPV.

In Figure 5.26 all four external command terminals are used with a flow control EHPV. The four potentiometers are set to provide different voltage inputs, with the voltage increasing from pot 1 to pot 4. A timed switching device such as a programmable controller is used to sequence the switches and increment the drive motor speeds. Additional circuitry (not shown) could be activated by the same switching control to open valves to provide the appropriate slurry for each stage of the operation. The ramping function can be used to minimize system shock and flow surges as well as to provide time for flushing the coarser grit before the higher wheel speed begins.

Figure 5.27 shows a system that can be used to provide two forward speeds and one reverse speed for a hydraulic cylinder. Such a circuit could be used in a manufacturing process requiring a rapid advance to push a workpiece into the work place, followed by a slow feed of the work piece through the machining operation. The cylinder is then retracted as rapidly as possible to pick up another workpiece. Notice that in this cycle the single EHPV provides a degree of hydraulic control that would require at least two directional control valves and two flow control valves in conventional circuitry.

5.8   SUMMARY

Electrohydraulic proportional control valves (EHPVs) provide an intermediate step between the simple on-off solenoid valve and the sophisticated servovalve. EHPVs combine directional, speed, and acceleration control into a single valve with an electronic controller. Although EHPVs are usually considered to be directional control valves, they can also be used for pure flow control (or throttling) or pressure control.

The performance of new-generation EHPVs rivals that of the lower level of servovalves, although they use less sophisticated electronics and seldom use load-related feedback. They are much less expensive than servovalves, even though some EHPVs have performance capabilities approaching those of servovalves. In many cases, therefore, it may be more cost effective to use an EHPV instead of a servovalve, especially if feedback is not needed.

In the next chapter we will explore servovalves and see how they differ from EHPVs, both mechanically and electronically.

REVIEW QUESTIONS

General

1. What is the functional difference between a standard solenoid and a proportional solenoid?
2. Explain the concept of operation of a proportional solenoid.

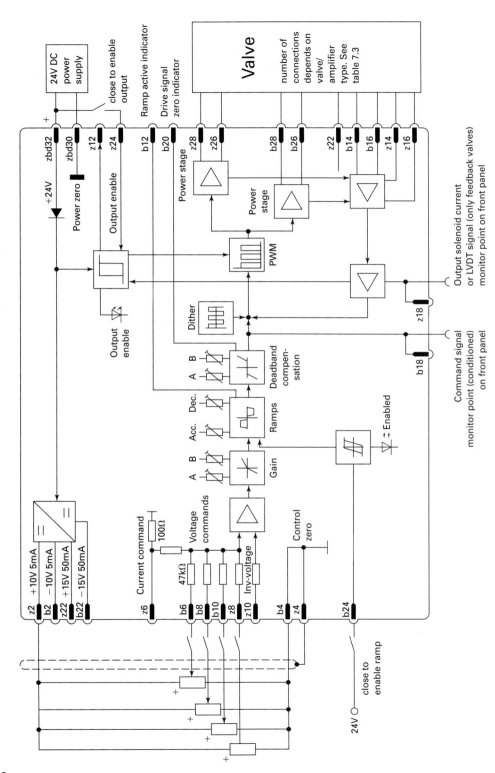

Figure 5.26 Wiring connections for four forward speeds.

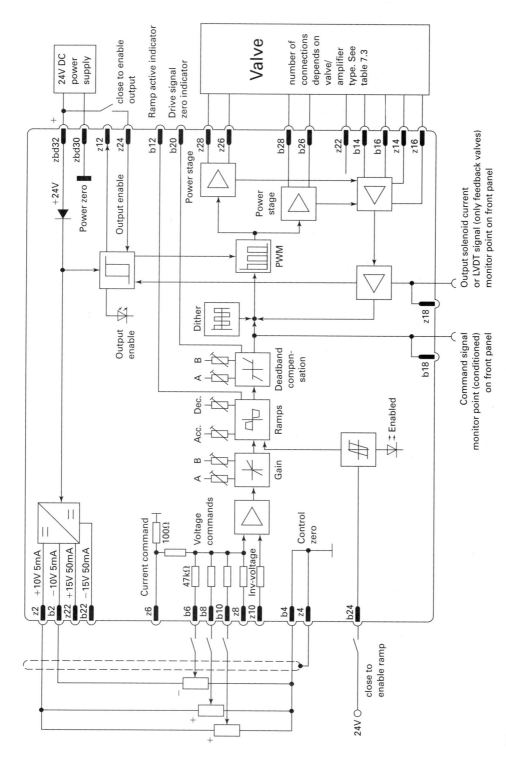

Figure 5.27 Wiring connections for two forward speeds, one reverse speed.

117

3. What is the benefit of the ramp function on a proportional controller?
4. What is the purpose of dither in a proportional circuit?
5. Explain the concept of pulse width modulation.
6. Explain the operation of an LVDT.
7. What is the purpose of the LVDT on some proportional control valves?
8. Explain the difference between force control and position control in proportional control valves.
9. Using standard solenoid valves and flow control valves, draw the hydraulic and electric control circuits that will provide four forward speeds and four reverse speeds for a hydraulic motor.

SUGGESTED ADDITIONAL READING

The Hydraulic Trainer, Vol. 2. 1986. Lohr am Main, Germany: Mannesmann Rexroth.
Industrial Hydraulics Technology. 1997. Cleveland, Ohio: Parker Hannifin Corp.
Electronically Controlled Proportional Valves. Tonyan, Michail J. 1985. New York: Dekker.
Vickers Industrial Hydraulics Manual. 1992. Rochester Hills, Mich.: Vickers, Inc.

CHAPTER 6

Servovalves

OBJECTIVES

When you have completed this chapter, you will be able to:

- Define a servovalve.
- Explain the differences between servovalves and proportional control valves.
- Describe and explain the operation of a torque motor.
- Explain the differences in and significance of overlapped, underlapped, and zero lapped spools.
- Describe the different types of hydraulic amplifiers and stages of servovalves and explain their operation.
- Explain the functions of an operational amplifier and discuss the various amplifier transfer functions for servovalve control.

6.1   INTRODUCTION

The ultimate in control is the ability to continuously monitor an output parameter and make instantaneous corrections to that parameter whenever it deviates from a predetermined set point. A simple example is the automobile driver who constantly watches the speedometer and makes changes to the force on the accelerator pedal when the speed varies from the posted speed limit. This situation represents *open-loop* control because it requires operator intervention. The use of a cruise control makes the same system *closed loop* because the corrections are made automatically without operator action once the initial command input is made.

   In machine motion control, servo systems involving continuous monitoring, feedback, and correction are used to improve efficiency, accuracy, and repeatability far beyond the capacity of a human operator. The most common and sophisticated applications of servovalves are in aerospace vehicles, particularly in primary flight controls. In aircraft, control surfaces such as ailerons, elevators, and rudders are positioned (and held in position) by servo units. In space vehicles, control during launch is provided by movable thrust nozzles that are positioned by servo units. Even the drill bits for angle drilling of oil wells are servo controlled. In this chapter we will look at servovalve theory and construction as well as applications.

6.2   DEFINITION OF A SERVOVALVE

In Chapter 5 we looked at electrohydraulic proportional valves. In the introduction to that chapter we attempted to determine the difference between EHPVs and servovalves. Because of recent advances in EHPV technology, those differences, especially in the area of performance, have narrowed so much that it is now easier to differentiate between the two by design and construction features rather than the traditional performance criteria. Table 6.1 summarizes these features.

As we look further into performance parameters (hystersis, deadband, and others), we will find other significant differences. To begin this chapter we consider the feedback feature to be the major differentiating factor. Although an EHPV system with some type of feedback from the main system will occasionally be found, a servovalve operating without feedback will seldom be found. In such cases the servovalve essentially becomes a very expensive EHPV.

A servomechanism is defined In *Merriam Webster's Collegiate Dictionary*, Tenth Ed., as "an automatic device for controlling large amounts of power by means of very small amounts of power and automatically [and continuously] correcting the performance of a mechanism." The automatic and continuous correction requires a return of information from the mechanism—feedback, in other words. Therefore, a servovalve operated without feedback is not a true servomechanism.

6.2.1 History of Electrohydraulic Servomechanisms

Maskrey and Thayer (1978) compiled a brief but very interesting history of these devices. Rather than offering another such document, we summarize their work here and refer you to their paper for a more complete treatment.

Table 6.1 Comparison of Servovalves and EHPVs

Feature	Servovalve	EHPV
Electrical operator	Torque motor	Proportional solenoid
Manufacturing precision	Extremely high	Moderately high
Feedback circuitry	Main systems as well as valve	Valve (depending on type), main system (seldom)
Cost (compared with solenoid valve)	Very expensive	Moderately expensive

The earliest recognized servomechanism is the water clock invented around 250 B.C. by the Alexandrian inventor Ktesbios. In his device, time was recorded by the level of water in a graduated vessel. Water flowed into this vessel at a controlled, constant rate from a water reservoir above it. The control of the flow rate from the reservoir involved a servomechanism.

As you may remember from your fluid mechanics courses, the velocity of flow from the outlet of a reservoir (or any vessel) is determined by the equation

$$v = \sqrt{2gh} \tag{6.1}$$

where v = velocity
   g = gravity
   h = height of water above the outlet

This relationship is known as Torricelli's theorem. The volume flow rate through that outlet depends on the size of the outlet and the fluid velocity. Thus,

$$Q = vA \tag{6.2}$$

From these equations we can see that as the water level in Ktesbios' reservoir went down the flow rate from the reservoir decreased. Consequently, the hours "got longer" (based on the use of a cylinder with equal graduations). Ktesbios's solution to this problem was to use a second reservoir mounted above the first. He used a float to modulate an orifice through which water was fed into the primary reservoir. This kept the water level (hence the flow rate) constant, resulting in hours of a constant length.

In addition to having scientific and technical value, this water clock was also a significant business and social achievement. Until its invention, time was kept primarily by sundials, which have two obvious drawbacks: It was not possible to keep time at night (or heavily overcast days, for that matter), and the length of the hour varied depending on the time of year. These problems were eliminated by the water clock (as long as somebody emptied the timing vessel at the appointed time). This device was used until the invention of the mechanical clock in the fourteenth century.

Numerous servomechanisms were invented during the Industrial Revolution in the mid-1700s, and afterward. Many were associated with steam boiler technology, where they were used to control water level, water and steam flow, steam pressure, and the speed and position of steam-operated mechanisms.

The technological advances of the early twentieth century—electric power, the automobile, and the airplane—spurred greater advances in servo technology. The first power steering unit with a mechanical feedback servo was invented in the late 1920s, although such units did not become popular until after World War II.

One of the few good things that result from any war is the advancement of technology. Immediately prior to World War II, Askania-Werke in Germany and Foxboro in the United States had begun to develop practical servomechanisms for controlling fluid-power circuits. Significant contributions were made during that time by Siemans of Germany and Tiebel in the United States. As a result of advancements in other fields (materials, fluids, electronics, control theory), some exciting progress was made immediately after the war. A chronology of these events follows:

1946—Tinsley of England: First two-stage valve

1947—Raytheon and Bell Aircraft: Two-stage valve with second-stage feedback (mechanical)

1947—Massachusetts Institute of Technology: True torque motor driver, two-stage valve with electrical second-stage feedback

1950—W. C. Moog Jr.: Frictionless first stage (flapper nozzle)

1953—T. H. Carson: Frictionless first stage with mechanical feedback from the second stage

1953—W. C. Moog Jr.: Symmetrical, double-nozzle bridge

1953—Wolpin: Isolation of torque motor from fluid

1957—R. Atchley: Jet pipe first stage

It is interesting to note that although there have been advancements in electronic components and circuitry, materials, and manufacturing processes, as well as a proliferation of applications throughout almost every phase of manufacturing, transportation, and the military, the servovalves used in fluid power systems today are of the same basic designs developed prior to 1960. There have been some evolutionary changes—miniaturization for aerospace applications, for instance—but no new, major revolutionary advances. Most advances have been in the areas of electronics, feedback transducers, and computer controls, all of which can be lumped under the heading of control technology.

6.2.2 Electrohydraulic Servomechanism Concepts

Figure 6.1 represents a typical fluid power system using a proportional valve to control the speed of a hydraulic motor. The EHPV is set to provide the necessary flow to drive the motor at the required speed. As long as there are no disturbances, the speed will remain constant. If, however, there is any change in the operating parameters—load, fluid temperature, viscosity, wear, and the like—the motor speed is likely to change. There is nothing in the system designed to detect that change and present the information to the valve controller to automatically correct the change and return the speed to the required level. Speed corrections are the responsibility of the operator, who must make the required control adjustments. Although this type of circuit is perfectly satisfactory for a very large number of applications, some require automatic and continuous corrections. These circuits require servomechanisms. (From this point on, we will simply refer to these mechanisms as servovalves.)

Figure 6.2 shows a circuit that has the same purpose as that of Figure 6.1, but in this circuit the operator has been relieved of responsibility for speed corrections. Instead,

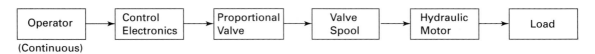

Figure 6.1 Proportional valve block diagram. Motor control circuit provides no correction for changes in motor rpm due to load changes or other factors.

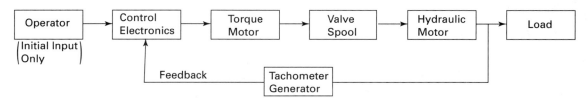

Figure 6.2   Servo control provides automatic and continuous corrections for any changes in motor rpm.

a tachometer generator has been installed that senses the load speed. This information is automatically and continuously fed back to the control electronics (usually a printed circuit board), where it is compared with the operator command signal input. If any difference is found between these two signals, the electronic circuitry automatically generates a correction signal proportional to the difference. That signal repositions the valve to correct the flow rate as required. This "sense and correct" function is continuous, so any and every change in load speed is automatically corrected. The system required to perform this function includes three major segments: the servovalve, the command electronics, and the feedback transducer. In the next sections, we will take a detailed look at each of these segments.

6.3   SERVOVALVES

Servovalves can be used in virtually any aspect of fluid power system operations, including:

a. positioning of cylinders and rotary actuators
b. speed of cylinders and motors
c. cylinder force and motor torque
d. acceleration and deceleration
e. system pressure
f. flow rate

The most common applications are for cylinder positioning and motor speed control. The valves for these functions incorporate both direction and flow control in a sliding spool arrangement that is positioned by a torque motor.

6.3.1 Torque Motors

A *torque motor* is illustrated in Figure 6.3. It is a rather simple electromagnetic device consisting of one or two permanent magnets, two pole pieces, a ferromagnetic armature, and two coils. The permanent magnets polarize the upper and lower polepieces, so that they present equal and opposite magnetic fields. Torque motors are very low power devices operated on low-voltage DC power.

Figure 6.4 demonstrates the concept of torque motor operation. The armature is mounted at its midpoint so that it is free to rotate through a very limited arc either clockwise or counterclockwise. The ends of the armature extend into the gaps between the

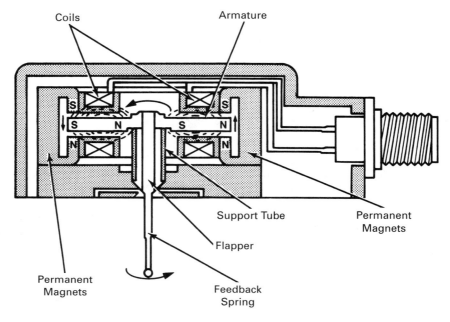

Figure 6.3 Servovalve torque motor. (Courtesy of Vickers, Inc.)

polepieces. The magnetic fields hold the armature in a neutral position. The two coils surround the arms of the armature to form two small electromagnets. When a current is passed through the coils, a magnetic field is generated. The polarity of the field depends on the direction of the current flow. In Figure 6.4 the current flow has caused the left end to become the south pole and the right end to become the north pole, resulting in a

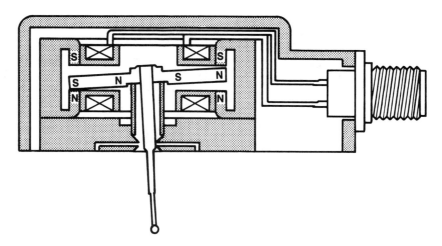

Figure 6.4 Servovalve torque motor operation. (Courtesy of Vickers, Inc.)

counterclockwise rotation of the armature. (We will see the effect of this rotation on valve operation a little later.)

The two coils of a torque motor may be connected in three different configurations—parallel, series, and the so-called push-pull arrangement. These options are illustrated in Figure 6.5. The push-pull arrangement is the most common. In the arrangement, leads B and D are both connected to ground through the control circuit amplifier. Leads A and C are connected to separate output terminals on the command amplifier. When the voltage

Figure 6.5 Torque motor coils can be connected in several combinations. Each causes a different operation.

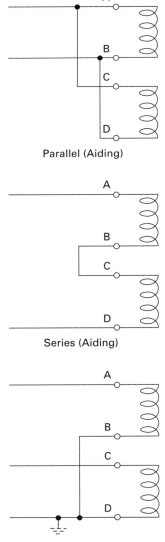

Parallel (Aiding)

Series (Aiding)

Push-Pull (Opposing)

input to both coils is equal, the armature is centered. Increasing the voltage input to one coil, while simultaneously reducing the input to the other coil by the same amount, causes the armature to rotate. The voltage can be varied from zero to its maximum value for each coil, but the polarity is never reversed. This means that the position of the armature is determined by differential torque. When the voltage is the same to both coils, the torque is equal, and the armature is centered. Any change in voltage to either coil results in rotation of the armature.

This push-pull method of connecting the coils is preferred for at least three reasons: First, any changes in current as a result of voltage fluctuations, temperature changes, or other causes are canceled by the equal and opposite effects on the coils. Second, there is more stability in armature positioning because of the opposing torque. Third, the power consumption is lower than for the other two circuits.

The input to this arrangement is expressed as a differential current, ΔI. This is the difference between the two coil currents. The control power is calculated from Equation 6.3:

$$P = (\Delta I)^2 R \qquad (6.3)$$

where P = control power
ΔI = differential current = $I_{AD} - I_{BC}$
R = resistance of one coil

Example 6.1: A torque motor is connected in a push-pull circuit. Each coil has a resistance of 20 ohms and is rated at 200 mA.
Find:

   a. The voltage of each coil when the armature is centered.
   b. The maximum value of ΔI.
   c. The maximum control power for the torque motor.

Solution:

   a. The maximum voltage for the coils is $E = I \times R$

$$E = 200 \text{ mA} \times 20 \text{ }\Omega = 4 \text{ V}$$

The armature is centered when

$$E = \frac{E_{max}}{2} = 2 \text{ V}$$

   b. The differential current is $I = I_{AD} - I_{BC}$
The maximum value will occur when the maximum voltage is applied to one coil (say, AD), so that zero voltage is applied to the other. In this case

$$\Delta I = I_{AD} - I_{BC} = \frac{E_{max}}{R} - 0 = \frac{4 \text{ V}}{20 \text{ }\Omega} = 200 \text{ mA}$$

   c. The maximum control power is then

$$P = (\Delta I)^2 R = (200 \text{ mA})^2 \, (20 \text{ }\Omega) = 0.8 \text{ W}$$

In the parallel connection the direction of rotation depends on the polarity of the input signal. Rather than opposing each other (as in the push-pull circuit) the coils in the parallel circuit assist each other. That is, they both attempt to move either clockwise or counterclockwise. Reversing the polarity reverses the direction of rotation. The control power is then found from

$$P = (I_P)^2 \left(\frac{R}{2}\right)$$ (6.4)

Where I_P = the total current through the circuit
         R = the resistance of each coil

Example 6.2: Repeat Example 6.1 for a parallel circuit connection.

Solution: The voltage to each coil will remain the same (4 V); however, the current through the circuit will increase because of the lower resistance. For a parallel circuit made up of two equal resistors, the equivalent resistance is $R/2$; in this case, 10 ohms.
    The value of I is the total current, which we find from

$$I_P = \frac{E}{R} = \frac{4\ V}{10\ \Omega} = 0.4\ A = 400\ mA$$

Finally, we find the control power from Equation 6.4:

$$P = (I_P)^2 \left(\frac{R}{2}\right) = (400\ mA)^2 \left(\frac{20\ \Omega}{2}\right) = 1.6\ W$$

In the series circuit the coils are assisting, rather than opposing, the armature rotation. As with the parallel circuit, a polarity change is required to change the direction of rotation. The control power for a series circuit is

$$P = (I_S)^2 (2R)$$ (6.5)

where I_S = the current in the series circuit
         R = the resistance of each coil

Example 6.3: Repeat Example 6.1 for a series circuit.

Solution: Here, the total resistance is $2R$, or 40 ohms. The maximum current will be 200 mA because of the series connection. The maximum voltage, then, is

$$E = IR = (200\ mA)(40\ \Omega) = 8\ V$$

The control power, from Equation 6.5, is

$$P = (I_S)^2 (2R) = (200\ mA)^2 (2)(20\ \Omega) = 1.6\ W$$

Notice that the series and parallel circuits have the same maximum power requirement, which is exactly twice that of the push-pull circuit.

It is interesting to note that these low-power torque motors can control a two- or three-stage valve that may be flowing 100 gpm or more at 2000 to 3000 psi. Taking the lower of these values, we see that the power output of the valve approaches 90,000 W. If we define power gain of the valve as the output power divided by the control power, we have

$$\text{Power gain} = \frac{90,000 \text{ W}}{1.6 \text{ W}} = 5.625 \times 10^4$$

which fits *Webster's* definition of a servomechanism.

6.3.2 Valve Spools

At first glance the hardware of a servovalve looks very similar to that of any spool-type directional control valve, that is, a sliding spool that operates in the bore of the valve body to open and close the flow paths between ports. The actual differences are found more in the manufacturing process and clearance specifications than in the basic design.

The servovalve spool and the bore in which it moves are very high precision components. Typically, spool and bore straightness and diametrical tolerances are held to ± 0.000050 in. The radial clearance between the spool and the bore is typically 3 to 5 μm (1 μm = 0.000039 in.). In most valves the radial expansion that results from holding the spool in your hand for a few minutes would prevent its insertion into the bore. To achieve this precision, a great deal of hand finishing is involved in the manufacturing process. The spool and body are quite often a matched set, and parts are not interchangeable.

Special spool surface finishes are often employed. Nitriding is often used to provide extra surface hardness and a glasslike smooth finish. This reduces friction and improves wear characteristics. In tests conducted by the Schenk Pegasus Corporation in 1982, nitrided and nonnitrided spools were run to 101 million cycles. The nitrided spools showed no change in pressure gain (which we will discuss later), whereas the nonnitrided spools had a 50% change, indicating a much higher degree of wear in nonnitrided units.

Servovalves may be either three- or four-way. The spools may have two, three, or four lands, depending on the function and on the manufacturer's preference. It has been shown that four-land spools can have slightly larger clearances without incurring unacceptable leakage. This means that they have improved wear characteristics and are somewhat more tolerant of contaminants in the fluid. The two outer lands also assist in keeping the spool precisely centered.

As with most spool-type valves, circumferential grooves are machined into the spool lands. The purpose of the grooves is to reduce side loading by equalizing the pressure around the spool and holding it centered in the bore. A spool with three grooves can have as little as 6% of the side force found in ungrooved spools.

Spool "lap" defines the width of the lands relative to the width of the ports in the valve bore. There are three possible lap configurations—overlap, underlap, and line-to-line. These are shown in Figure 6.6.

By far the most common condition is the line-to-line (or *zero overlap*) spool. Here, the land width exactly matches the port width. Thus, when the spool is centered, there is

Figure 6.6 Spool lap configurations.

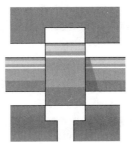

(a) Line-to-line or Zero Overlap Spool

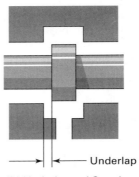

Underlap

(b) Underlapped Spool

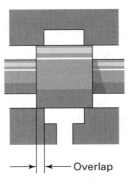

Overlap

(c) Overlapped Spool

no flow. Any movement of the spool—regardless of how little—results in flow through the valve. This valve is suitable for closed-loop position, speed, and force control applications because of its precise metering characteristics about the null (neutral) position. Unfortunately, even a small amount of wear on either the land or the port edge will result in leakage in the null position.

Overlapped spools have lands that are 0.5 to 5% wider than the ports. This spool has the advantage of providing lower leakage flow in the null position than the line-to-line configuration; however, the overlap means that the precision achievable about the null position is compromised because of the relatively large deadband. For instance, when used as a position controller, a cylinder that is being extended will stop at a different position when being retracted, even with the same command input. An overlapped valve can be satisfactorily employed as a speed controller as long as it is operated well away from its null position.

In many servovalve control circuits, *dither* is used to reduce the effects of static friction (termed *stiction*). Dither is a very low amplitude command signal superimposed over the normal command signal that results in a continuous, very short stroke, lateral oscillation of the spool. In such systems, a slight overlap may be used to prevent unacceptable leakage in the null position.

An underlapped spool has lands that are 0.5 to 1.5% narrower than the ports. This design is often referred to as "open center," although there really are no open-center servovalves. The underlap is far too small to be a true open center. This type valve provides very rapid response to commands about the null position, but it has the disadvantage of having nonlinear flow characteristics near null. This compromises control to some extent.

6.3.3 Valve Configurations

Servovalves may be single-stage (also called direct-acting), two-stage, or three-stage, depending primarily on the flow requirements of the system.

Single-stage valves may be used where the flow requirements are low (usually less than 5 gpm, depending on the valve design). These valves commonly utilize a sliding spool mechanically connected to the torque motor armature. The flow capacity is dictated by the low force available from the torque motor and the limited stroke of the spool.

Figure 6.7 shows a single-stage servovalve. The mechanical connection between the torque motor armature and the spool is a stiff wire. When there is no command input to the torque motor, the armature is in its neutral (nulled) position, which, in turn, causes the spool to be in the nulled position, and there is no flow through the valve. A clockwise deflection of the armature pushes the spool to the left, opening up flow paths from P to B and A to T. A counterclockwise deflection opens P to A and B to T.

For higher flow rates, two- or even three-stage valves must be used. In these valves the second and third stages are always sliding spools that are pilot operated from the previous stages. The first stage may use a sliding spool, but there are other designs, also. We will look at some of these.

Since we have already discussed the operation of a single-stage sliding spool valve, let's start with the spool-type pilot stage. Figure 6.8 shows such a valve being used to control the direction and speed of a hydraulic motor. We see that the pilot spool is positioned by the torque motor, as before. There is a major difference in the pilot section, however, that is not in the valve we looked at earlier—a pilot stage sleeve. This sleeve is associated with the internal feedback mechanism of the valve, which we will discuss shortly. Notice also that there are two pressure sources—p_c is the control pressure for piloting the main spool, and p_s is the supply pressure for operating the system.

In the neutral position, shown in Figure 6.8, the middle spool land blocks off the pilot port to the left (large) end of the main (second-stage) spool. The pressure on the large end of the spool is half the control pressure. The pressure at the small end of the spool is always equal to the control pressure. In order to be balanced, then, the large end

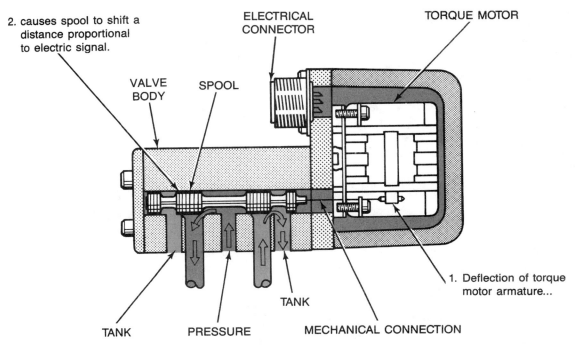

2. causes spool to shift a
   distance proportional
   to electric signal.

ELECTRICAL
CONNECTOR

TORQUE MOTOR

VALVE
BODY

SPOOL

1. Deflection of torque
   motor armature...

TANK

TANK

PRESSURE

MECHANICAL CONNECTION

Figure 6.7 Single-stage spool-type servovalve is shifted directly. (Courtesy of Vickers, Inc.)

must have twice the area of the small end. The result is that the main spool is static. Because no input has been made to the pilot spool, the main spool is in the neutral position, so there is no flow to the hydraulic motor.

A counterclockwise movement of the torque motor armature pushes the pilot spool to the left. This opens up the port (through the pilot stage sleeve) to the large end of the main spool. The pressure in that end increases, causing a force imbalance that shifts the main spool to the right and opens flow paths P to A and B to T.

This movement of the main spool also initiates the internal feedback action. As the main spool moves to the right it pushes in the feedback linkage. This linkage transmits the main spool movement to the pilot stage sleeve, causing the sleeve to slide to the left. When the sleeve port moves to the point where it is blocked by the pilot spool land, flow to the large end of the main spool stops. Consequently, the main spool stops. The exact stopping point is predetermined and is controlled by the deflection of the torque motor. Because the opening created by the movement of the main spool constitutes a control orifice, both the direction and speed of rotation of the hydraulic motor are set.

Deenergizing the torque motor returns the pilot spool to its neutral position, which opens up a flow path through the pilot stage sleeve and ports the large end of the main spool to the tank. The resulting force imbalance shifts the spool to the left. This allows the pilot stage spring to push the spool back to the right. When the sleeve port realigns with the pilot spool land, all pilot flow stops; the main spool stops (in its neutral position), and the hydraulic motor stops. A similar, but opposite, sequence takes place when

2. In neutral, large
pilot end is blocked
at pilot valve in the
static condition. This
pressure = 1/2 control
pressure (p_c).

3. Control pressure
is present here and
at small end of
mainspool.

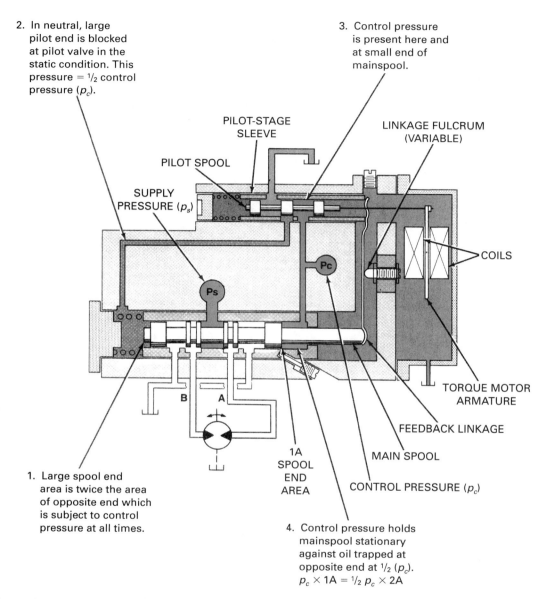

PILOT-STAGE
SLEEVE

LINKAGE FULCRUM
(VARIABLE)

PILOT SPOOL

SUPPLY
PRESSURE (p_s)

COILS

Pc

Ps

B   A

TORQUE MOTOR
ARMATURE

FEEDBACK LINKAGE

1. Large spool end
area is twice the area
of opposite end which
is subject to control
pressure at all times.

1A
SPOOL
END
AREA

MAIN SPOOL

CONTROL PRESSURE (p_c)

4. Control pressure holds
mainspool stationary
against oil trapped at
opposite end at 1/2 (p_c).
$p_c \times 1A = 1/2 \, p_c \times 2A$

Figure 6.8 Two-stage servovalve is pilot-operated. (Courtesy of Vickers, Inc.)

the torque motor armature is deflected clockwise. I suggest you step through the sequence on your own.

Figure 6.9 shows a double flapper nozzle pilot stage. The flapper is a physical part of the torque motor armature and extends into a fluid cavity inside the valve. Control pressure for piloting the main spool enters the pilot section through ports on opposite sides of the flapper. The pressure is then channeled to nozzles located in the fluid cavity into

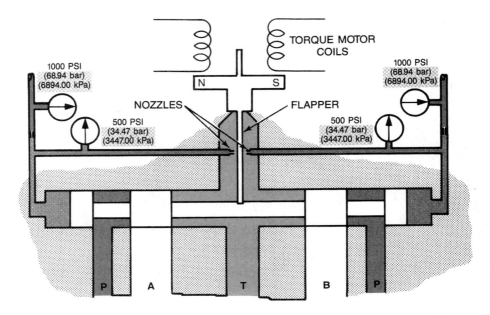

Figure 6.9 Flapper-nozzle-type servovalve. (Courtesy of Vickers, Inc.)

which the flapper extends. These channels are also connected to the pilot chambers on the ends of the main spool.

When the torque motor is not energized, the armature is centered, which positions the flapper exactly in the middle of the cavity. Because the distances between the nozzle lands and the flapper are very small, this, in effect, forms small orifices that restrict the flow and generate a pilot pressure. With the flapper exactly centered, both orifices are the same size, resulting in the same pressure in both pilot chambers. This causes a force balance on the main spool and holds it in its neutral position.

If the armature is deflected counterclockwise, the flapper moves toward the right-hand nozzle. This movement reduces the orifice associated with that nozzle while increasing the orifice for the opposite nozzle. This increases the pressure in the right-hand pilot chamber and decreases the opposite pressure. The result is a force imbalance that shifts the main spool to the left and opens the related flow paths.

The internal feedback mechanism in the flapper-nozzle valve is a flexible metal rod (usually termed a spring) that is attached to the end of the flapper and inserted into a ball joint in the main spool. As the main spool shifts to the left the feedback spring exerts a force on the flapper (and, consequently, the armature), which tends to return it to its neutral position. The force is proportional to the distance the spool moves, so as the spool shifts to the left the restoring force in the flapper increases. When the force is sufficient to center the flapper, the control orifices are again equal, as are the pressures in the pilot chambers, so the main spool stops. It remains in that position until a subsequent command signal causes the armature to move again.

A variation of the double flapper-nozzle first stage is the single flapper nozzle. As you might expect, the concept employed is the same, except there is only one nozzle. The

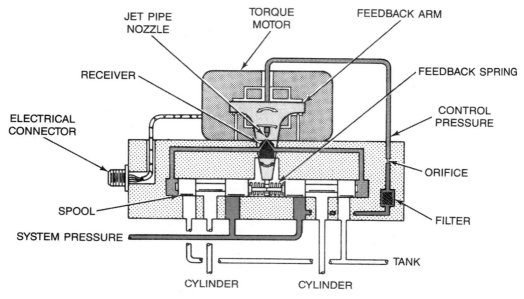

Figure 6.10 Jet pipe servovalve. (Courtesy of Vickers, Inc.)

pressure that occurs in the associated pilot chamber works against a calibrated spring on the opposite end of the main spool.

Another pilot stage is the jet pipe stage shown in Figure 6.10. In this device the jet nozzle is attached to, and moves with, the torque motor armature. The pilot fluid is directed by the jet nozzle into two receiver ports that connect to the pilot chambers at the ends of the main spool. In the neutral position the fluid jet is evenly divided, resulting in equal pressures in the pilot chambers. As a result, the main spool remains stationary. A deflection of the torque motor directs the jet more directly into one of the receiver ports, causing a force imbalance and main spool movement. Internal feedback is provided by a feedback spring in a manner similar to that in the flapper-nozzle system.

A variation of the jet pipe concept is the deflector jet valve shown in Figure 6.11. In this valve the jet pipe does not move. Rather, deflectors are moved into the fluid jet, which direct it into the appropriate receiver ports.

The standard symbols for servovalves are shown in Figure 6.12. The symbols give no hint as to the number of stages or type of pilot stage employed. As with all graphic symbols, they show function only. The directional/flow control servovalves are always three-position, infinitely positionable units and usually have closed centers. The actuator symbol represents the summing junction of the electronic amplifier, which receives two input signals (reference and feedback) and produces a single output signal. We will discuss the electronics later in this chapter.

6.3.4 Pressure-Flow Characteristics

Servovalves are generally considered to be "high-pressure" devices. They are usually pressure rated at 3000 psi, although most are capable of operating at 5000 psi or higher. Servovalves

Figure 6.11 Deflector jet
valve.

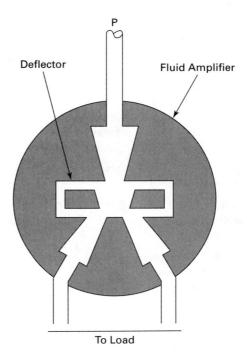

are typically flow rated at 1000 psi differential; that is, the flow rate stated for the valve is
the flow that occurs at a 1000 psi pressure drop across a fully open valve. There is a very deliberate logic in the choice of 1000 psi. The majority of servo systems use a working pressure of 3000 psi. It can be shown mathematically that the maximum transmission of power
from the pump to the cylinder or motor occurs when the pressure drop through the valve is
one-third of the system working pressure. This parameter is related to the basic orifice equation, which describes the flow rate through an orifice. Thus, the valves are flow rated at the
maximum power transfer pressure drop based on the popular 3000 psi system. It has also
been determined that valve control is optimized at the 1000 psi pressure drop.

Figure 6.12 Servovalve
symbols.

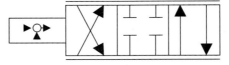

(a) Line-to-line and overlapped spool

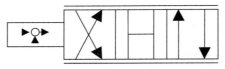

(b) Underlapped spool

If the pressure drop across the valve is different from the optimum, the flow rate will change proportionately. The nonoptimum flow rate can be calculated using Equation 6.6:

$$Q = Q_R \sqrt{\frac{\Delta p}{\Delta p_R}} \tag{6.6}$$

where Q = adjusted flow rate
$\quad\quad Q_R$ = rated flow
$\quad\quad \Delta p$ = actual pressure drop
$\quad\quad \Delta p_R$ = rated pressure drop (usually 1000 psi)

Example 6.4: A servovalve is flow rated at 15 gpm at 1000 psi differential. It is to be operated in a 2000 psi system. What will its adjusted flow rate be at the optimum power transfer Δp?

Solution: For optimum power transfer, the pressure drop across the valve should be 2000/3 = 667 psi.
From Equation 6.6, we see that

$$Q = Q_R \sqrt{\frac{\Delta p}{\Delta p_R}} = 15 \sqrt{\frac{667}{1000}} = 12.25 \text{ gpm}$$

The term "1000 psi pressure drop" seems to imply that the drop occurs in one flow path through the valve. In fact, the pressure drop in a properly sized valve will occur in two 500 psi stages—one going to the actuator, the other returning from the actuator, as illustrated in Figure 6.13. The servovalve is shown actuated in the drawing for clarity.

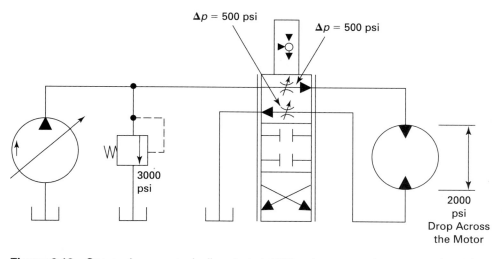

Figure 6.13 Servovalves are typically rated at 1000 psi pressure drop across the valve in a 3000 psi system.

With the pump compensator set at 3000 psi, the pressure is reduced to 2500 psi across the P to A path through the valve. If we assume the motor is operating at full load, 2000 psi is dropped across the motor. The final 500 psi is then dropped through the B to T side of the valve.

If the motor is not fully loaded, then system pressure drops below 3000 psi. In this case the 1000 psi drop across the valve still occurs, but the power transfer diminishes.

If the motor is overloaded so that more than 2000 psi is required to rotate it, then the 1000 psi drop across the valve is not possible. The result is a lower flow rate (Equation 6.6 applies), so the motor will turn more slowly. The extreme of this condition is a stalled motor. In this case, the full 3000 psi will be dropped across the motor. Obviously, there is no flow in this case.

Because the servovalve represents a major pressure drop, it will also generate a considerable amount of heat, which can be calculated from Equation 6.7:

$$\text{HGR} = \text{HP} \times 42.4 \text{ Btu/min/HP} \tag{6.7}$$

where   HGR = heat generation rate
        HP = horsepower loss across the valve

$$\text{HP} = \frac{\Delta p \times Q}{1714} \tag{6.8}$$

The fluid temperature rise across the valve can be found using Equation 6.9:

$$\Delta T = \frac{\text{HGR}}{C_p W} \tag{6.9}$$

where

ΔT = temperature rise
C_p = specific heat in Btu/lb·°F
W = weight flow rate = γQ
γ = fluid specific weight

(*Note*: Equation 6.9 lacks mathematical rigor because specific heat is correctly expressed in terms of mass rather than weight. The use of weight flow rate rather than mass flow rate corrects this discrepancy.)

As a rule of thumb, we can estimate the temperature rise by Equation 6.10:

$$\Delta T = \frac{0.75 \text{ °F}}{100 \text{ psi}} \times \Delta p \tag{6.10}$$

provided that C_p is between 0.45 and 0.5 Btu/(lb·°F), and the fluid has a specific gravity between 0.85 and 0.9.

Example 6.5: For the servovalve of Example 6.4, determine the horsepower loss, the heat generation rate, and the temperature rise across the valve. The fluid has a specific heat of 0.47 Btu/(lb·°F) and a specific weight of 54.9 lb/ft^3.

Solution: The horsepower loss (from Equation 6.8) is

$$\mathrm{HP} = \frac{\Delta p \times Q}{1714} = \frac{(667)(12.25)}{1714} = 4.77 \text{ HP}$$

The resultant heat generation rate is found from Equation 6.7 to be

$$\mathrm{HGR} = \mathrm{HP} \times 42.4 \text{ Btu/min/HP}$$

$$= (4.77)(42.4 \text{ Btu/min})$$

$$= 202.1 \text{ Btu/min}$$

The increase in the fluid temperature is calculated using Equation 6.9:

$$\Delta T = \frac{\mathrm{HGR}}{C_p W}$$

The weight flow rate is

$$W = \gamma Q$$

$$= (54.9 \text{ lb/ft}^3)(12.25 \text{ gal/min})(\text{ft}^3/7.48 \text{ gal})$$

$$= 89.9 \text{ lb/min}$$

Therefore,

$$\Delta T = \frac{202.1 \text{ Btu/min}}{(0.47 \text{ Btu/(lb·°F))}(89.9 \text{ lb/min})}$$

$$= 4.8 \text{ °F}$$

Notice that no flow rate is implied in the calculation. This is the temperature rise experienced by every drop of oil that flows through the valve. The rise occurs in the time required for the fluid to flow through the portion of the valve where the pressure drop occurs—generally, a few milliseconds.

Although the weight flow rate appears in the equation, the temperature rise is actually independent of flow rate. The heat generation rate in the numerator is a function of horsepower which, in turn, is a function of flow rate. Thus, the Q in the numerator cancels the Q (or weight flow rate) in the denominator, which is why the estimation shown in Equation 6.10 does not include a flow rate term. This means, simply, that the temperature rise for any given fluid is a function of the pressure drop only.

In this example, that estimated temperature rise is

$$\Delta T = \frac{0.75 \text{ °F}}{100 \text{ psi}} \times 667 \text{ psi} = 5.0 \text{ °F}$$

a reasonably close estimate.

6.3.5 Valve Performance

The performance of a servovalve can be described by numerous parameters. These are generally divided into two categories—dynamic response and static response. The dy-

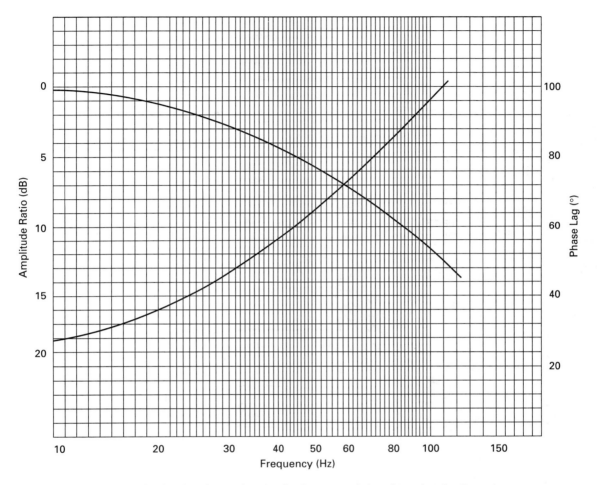

Figure 6.14 A typical Bode plot shows the amplitude rate and the phase lag for the valve.

namic response characteristics are the same as those for proportional valves—frequency response and amplitude ratio. Figure 6.14 is a Bode diagram showing these characteristics for a specific valve. (Bode diagrams are discussed in detail in Chapter 7.)

During the normal operation of a valve, it is likely to experience the same current input frequently. Sometimes this specific current input (we will call it a set point) is approached from a lower current setting. At other times, it is approached from a higher setting. All valves have the characteristic that this current set point will result in different spool positions, depending on whether it is approached from a lower or higher current. This characteristic is termed *hysteresis*. Typical hysteresis curves for servovalves and proportional control valves are shown in Figure 6.15. Hysteresis is expressed as the percent difference in the rated current required to give the same output when approached from higher and lower inputs. For servovalves it is typically 1 to 2%. To overcome the problem

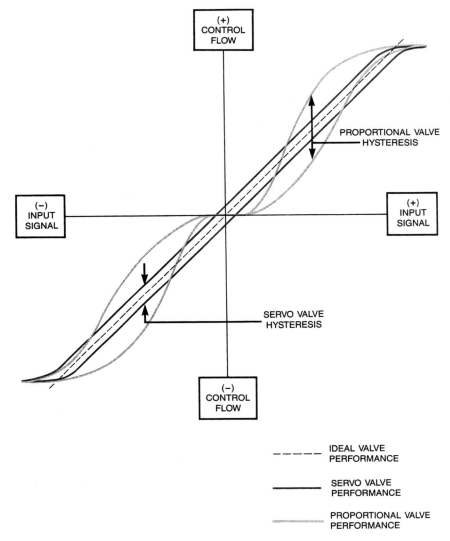

Figure 6.15 Hysteresis for a servovalve and a typical proportional valve. (Courtesy of Vickers, Inc.)

of hysteresis, some controllers are designed so that the set point is always approached from the lower side. This requires a deliberate undershoot when approaching from the high side.

A second important valve characteristic is the valve *deadband*. Deadband occurs only at the null position, as shown in Figure 6.16. It defines the current required to move the spool from the exact centered position to the position where the first flow output is seen. It is usually expressed in milliamps or percent rated current. Deadband is the result

Figure 6.16 Deadband occurs only at the null position.

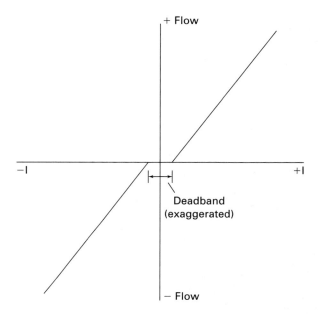

of spool inertia, overlap, static friction, and any other forces that might impede the initial motion.

A similar phenomenon is *threshold*. Threshold current is the smallest input current required to overcome spool inertia and other impeding forces to cause the spool to move. The primary difference between threshold and deadband is that threshold occurs throughout the spool stroke, whereas deadband occurs only at the null position. Threshold contributes to the deadband rating.

Information concerning hysteresis, deadband, and other valve performance characteristics is presented in the valve specification sheets available from the valve manufacturer. These characteristics can be significant in evaluating the suitability of a valve for a specific application.

6.4   ELECTRONICS

As we have seen, the servovalve itself is a very sophisticated electromechanical device. It is only one portion of the total system, however. In order for the valve to perform to its potential, some fairly sophisticated control electronics are required.

A block diagram of a typical control circuit for a servovalve is shown in Figure 6.17. All components involved in servovalve control (including the torque motor itself) operate on low-voltage DC power. There are several reasons for this, including ease of regulation, low power requirements, low heat generation, and the absence of the high inrush current characteristic of AC power.

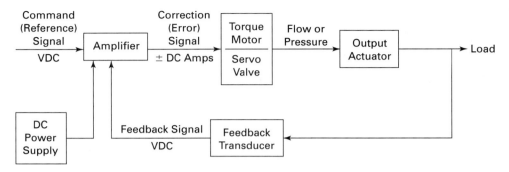

Figure 6.17 Block diagram of a closed-loop servo system.

The DC power supply may operate with either AC or DC as its input. Its output, however, must be a well-regulated DC signal with very little ripple permitted. Output voltage may be from 10 to 24 VDC, depending on the requirements of the particular system. In addition to powering the amplifier, the power supply may provide power for the input command and the feedback transducer.

The input command may come from many different devices. Although analog signals are most commonly used, digital signals are becoming more popular. Input devices include potentiometers, computers, programmable controllers, various types of switches, and manually operated joy sticks (which are, in fact, potentiometers). We will discuss amplifiers in some detail in the next section. Transducers will be addressed in Chapter 8.

6.4.1 Amplifiers

A basic amplifier is a relatively simple device that is designed to increase some aspect of an electrical input and produce an output that is proportional to the input. Most amplifiers are linear, which means that there is a constant proportionality of output to input. This constant of proportionality is termed the *gain* of the amplifier.

Although we are using the term "gain" here to indicate an increase, that is actually a very narrow definition. Basically, gain can be defined as the output divided by the input, or

$$\text{Gain} = \frac{\text{Output}}{\text{Input}} \tag{6.11}$$

This equation can apply to any output that results from any input. For instance, suppose you go to the local ice cream store. You give the clerk $2.39 (input) and the clerk gives you a banana split (output). The "gain" is

$$\text{Gain} = \frac{\text{Output}}{\text{Input}} = \frac{\text{Banana Split}}{\$2.39}$$

(There may be a "gain" of weight, also, and that *is* a valid use of the term.)

On a more technical level, we know that the speed of a hydraulic motor is a function of the flow rate going through it. Thus, if we have a motor that turns at 1600

Figure 6.18 Schematic for an NPN transistor.

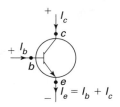

rpm (output) when the flow rate is 10 gpm (input), we can say that the gain of the motor is

$$\text{Gain} = \frac{\text{Output}}{\text{Input}} = \frac{1600 \text{ rpm}}{10 \text{ gpm}} = 160 \text{ rpm/gpm}$$

Returning to the amplifier, we will use the narrow definition and assume that a gain is an increase. If we have an amplifier that has an output of 10 V for a 2 V input, its gain is

$$\text{Gain} = \frac{10 \text{ V}}{2 \text{ V}} = \frac{5 \text{ V}}{\text{V}}$$

The concept of gain, along with feedback, phase lag, and other performance characteristics will be discussed in much more detail in Chapter 7. These parameters will be considered for an entire system instead of just the valve and amplifier.

Most amplifier circuits use one or more transistors as their basic electronic element. A transistor is a semiconductor element that produces an output that is proportional to the input. Although the proportionality is constant, the actual output is controlled by a second input. Let's look at a simplified illustration.

Figure 6.18 shows the schematic symbol for a typical transistor. There are three connections on the transistor—a base (b), a collector (c), and an emitter (e). The current flow from the collector to the emitter will be proportional to the current flow from the base to the emitter. Table 6.2 shows the results of increasing the base current. Notice that the collector current is always 100 times greater than the base current. If we consider the collector current to be the output and the base current the input, we see that the gain is 100. Notice also that the emitter current is the sum of the collector and base currents, as we would expect (from Kirchhoff's law).

Table 6.2 Transistor Output versus Input

Ib (mA)	Ic (mA)	Ie (mA)
0	0	0
0.5	50	50.5
1	100	101
1.5	150	151.5

6.4.2 Operational Amplifier

Servovalve control circuitry employs a device known as an *operational amplifier*, or *op amp*. An op amp is a much more complex device than the basic amplifier we have been discussing. In fact, an op amp will likely have several amplifiers included in its circuitry to perform different operations on the control signal. It has the following basic functions in system operation:

a. Compares command (reference) and feedback signals
b. Provides command signals to the servovalve torque motor
c. Provides gain to increase the system sensitivity
d. Provides gain adjustments to optimize system performance
e. Protects the valve torque motor from electrical damage by limiting the current input.

Op amps are very high gain devices, with gain factors of 10^5 or more being possible (although such high gains are not used with hydraulic systems). They have high input impedance, so they draw very little current from the input, and they have low output impedance, so they can produce a high current output with relatively low voltage.

The high gain of an op amp makes it somewhat impractical for use as a basic amplifier that operates in an open mode. For example, suppose we have an amplifier connected as shown in Figure 6.19. With a ± 15 VDC power supply, the maximum output from the amp would be ± 15 V (the amp cannot "make" power). With a gain of 10^6, the maximum output voltage would occur with an input of only 0.000015 V, since

$$\text{Gain} = \frac{\text{Output}}{\text{Input}}$$

and
$$\text{Input} = \frac{\text{Output}}{\text{Gain}} = \frac{15 \text{ V}}{10^6} = 1.5 \times 10^{-5} \text{ V}$$

This is a ridiculously low voltage for any practical application, so this amplifier would seem to be of little use. An obvious solution would be to use less gain, but there is another solution, involving the use of a closed-loop circuit, that could prove to be more satisfactory in many applications.

Figure 6.19 An amplifier with a gain of 10^6 would require only 1.5×10^{-5} V to provide a 15 V output.

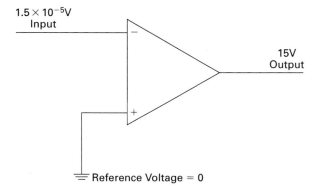

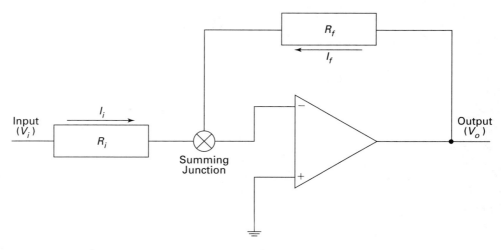

Figure 6.20 Op amp with voltage applied at input.

A closed loop requires a circuit such as that shown in Figure 6.20. In this circuit, R_i is a resistor placed in the input line, and R_f is in a feedback circuit connecting the amp output back to the input through a point called a *summing junction* or *comparator*. Because of the design of the amplifier, since the positive terminal is at 0 V (ground), the negative terminal will also be at 0 V.

From Kirchhoff's law, we know that

$$I_i + I_f = I_{\mathrm{amp}} \tag{6.12}$$

Because of the high input impedance of the amp, the current into it is small enough to be considered to be zero, thus,

$$I_i + I_f = 0$$

or

$$I_i = -I_f$$

From Ohm's law, we see that

$$I_i = \frac{V_i}{R_i}$$

and

$$I_f = \frac{V_f}{R_f}$$

Therefore, we have

$$\frac{V_i}{R_i} = \frac{-V_f}{R_f}$$

Solving for V_f gives us

$$V_f = \frac{-R_f}{R_i} V_i$$

But since V_f is the output voltage, V_o, we can write

$$V_o = \frac{-R_f}{R_i} V_i$$

This means that the gain of the closed loop circuit is equal to

$$\frac{-R_f}{R_i}$$

Example 6.5: The values of the resistors in Figure 6.20 are $R_i = 1000\ \Omega$ and $R_f = 2000\ \Omega$. Find the voltage output if the input voltage is $+10$ V.

Solution:

$$V_o = \frac{-R_f}{R_i} V_i = \frac{-2000\ \Omega}{1000\ \Omega} (10\ \text{V}) = -20\ \text{V}$$

We see that the gain of the amplifier is -2, or -2 V/V.

An obvious extension of this concept is to use variable resistors in place of R_i and R_f to allow more flexibility in the operation of the amplifier. This allows the operator to adjust the amplifier gain as necessary. Almost all op amps employ variable resistors in some configuration similar to Figure 6.21.

Thus far we have considered only the magnitude of the output from the op amp. Another factor that needs to be considered is the manner in which the amplifier achieves the output, which is often termed the *transfer function*. It is determined by the equa-

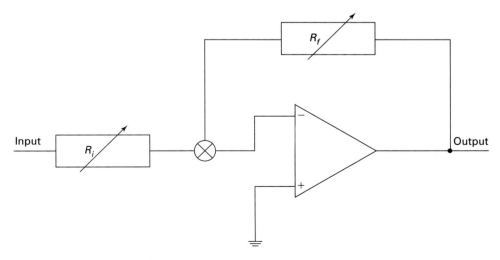

Figure 6.21 Op amp with variable resistors.

tions that describe the operation of the amplifier components. We will not analyze the equations but rather will look at the results of the transfer functions on the output. Although there are numerous possible transfer functions, they can be summarized in the six basic types shown in Figure 6.22. The graphs indicate the amplifier's response to a step input.

The *proportional (P) element* provides a step output that is proportional to the step input. Mathematically, this relationship can be described as

$$X_{out} = K \cdot X_{in} \qquad (6.13)$$

where K is the constant of proportionality, or gain, of the element and X is some unspecified parameter (voltage, current, flow, etc.).

A variation of the proportional element is the *first-order proportional* (PT1) *element*. Rather than providing a step response, the output is a first-order curve that has a rise rate determined by the time constant (T) of the element. Mathematically,

$$T \cdot \dot{X}_{out} + X_{out} = K \cdot X_{in} \qquad (6.14)$$

where $\dot{X}$ represents the first derivative of X.

A further variation is the *second-order proportional* (PT2) *element*, which reaches the final output value at a rate determined not only by the time constant (T) of the element but also by its damping function (D) and is expressed as

$$T \cdot \ddot{X}_{out} + 2DT\dot{X}_{out} + X_{out} = K \cdot X_{in} \qquad (6.15)$$

where $\ddot{X}$ represents the second derivative of X.

Notice that the right side of all three of these equations is $K \cdot X_{in}$. The output from all three elements is the same, but they arrive at that point by different paths.

The *integrating* (I) *element* produces a time-dependent output in response to a step input. For instance, a voltage step input produces a linear change of output with respect to time for as long as the voltage remains at its step value (that is, does not drop to zero). A peculiarity of this device is that the output does not fall back to zero when the input returns to zero. Rather, it remains at the value it had reached at that time. The transfer function for this element is

$$X_{out} = K \int X_{in} \, dt \qquad (6.16)$$

This type element is commonly used for speed control in fluid power systems.

A *time-delay* (TD) *element* is similar to the basic proportional element, except that the step increase in output occurs at some time delay after the step input.

The *differential* (D) *element* differs from the others in that its output depends on the time rate of change of the input rather than on the actual value of the input. Therefore, its response is based on a ramped input signal rather than a step input. For this reason, amplifiers normally use a derivative element in conjunction with P and/or I elements. Mathematically, the D element is represented by

$$X_{out} = K \cdot \dot{K}_{in} \qquad (6.17)$$

Most op amps used to control servovalves utilize a combination of these elements. The most common combinations are the PD controller for *positioning* systems and the PID controller for *velocity* systems.

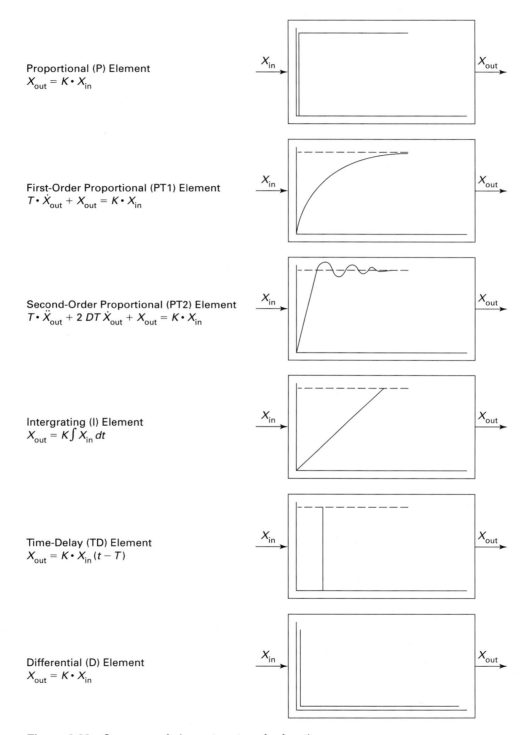

Proportional (P) Element
$X_{out} = K \cdot X_{in}$

First-Order Proportional (PT1) Element
$T \cdot \dot{X}_{out} + X_{out} = K \cdot X_{in}$

Second-Order Proportional (PT2) Element
$T \cdot \ddot{X}_{out} + 2\, DT\, \dot{X}_{out} + X_{out} = K \cdot X_{in}$

Intergrating (I) Element
$X_{out} = K \int X_{in}\, dt$

Time-Delay (TD) Element
$X_{out} = K \cdot X_{in}\,(t - T)$

Differential (D) Element
$X_{out} = K \cdot X_{in}$

Figure 6.22   Summary of elementary transfer functions.

Although our discussion in this section has been limited to the concept of electronic amplifiers, you have probably noticed that the equations presented can be used to describe many other phenomena. Relating these electronic concepts to others with which you may be more familiar may enhance your understanding of the electronics.

Figure 6.23 shows a typical servovalve control module containing an operational amplifier. This is a multifunction unit; therefore, it contains more circuitry than might normally be found on other control modules. Notice that there are three stages of amplification on the unit. Although three amplifier *stages* are shown, the module actually contains five amplifiers. As with the proportional valve control PCB we discussed in Chapter 5, each amplifier inverts the input signal polarity. Thus, a positive input results in a negative output and vice-versa.

Notice that there are also a number of potentiometers and switches. Let's look at the functions of these devices, beginning with the switches. Bear in mind that this block diagram is a very simplistic representation of the electronic componentry; therefore, the manner in which the switches affect the circuitry may not readily be apparent.

Switch S1 is closed when the module is used for controlling velocity, the speed of a hydraulic motor for example. Closing this switch connects a feedback capacitor into the circuitry and ties an integrating element into the voltage amplifier, giving a PID op amp. Recall that the peculiarity of an integrating element is that the output remains at the level it had reached when the input signal was removed. This means that the torque motor will continue to receive a signal that will hold it open at the point where the feedback and command signal were equal, resulting in a constant motor speed. (We will discuss this concept later in this chapter.) Switch S2 is open when S1 is closed.

If S2 is closed, and S1 is opened, a feedback resistor is selected instead of a feedback capacitor. This puts the preamp in a linear mode and makes the voltage amplifier a PD unit. This selection is used for positioning systems, because the servovalve must shut off flow to the actuator when the commanded position has been reached.

If both S1 and S2 are open, the voltage amplifier operates as an open-loop device. In this mode it can be used as a high-gain comparator with a gain of approximately 20,000.

If S3 is open, the voltage amplifier operates independently. Normally, this switch is closed, so that the power amplifier is connected to the power stage preamp. Switch S4 can be closed to bypass the 51 kΩ resistor for a special application.

The gain range of the voltage amplifier can be changed by manipulating S5. Closing this switch bypasses the 270 kΩ resistor, providing a high gain range.

Switch S6 adds a diode threshold limiter when closed. This limits the maximum error signal to reduce overshoot when using the integrating mode for velocity control. S5 must be open when S6 is closed.

The potentiometers (or pots) are used to fine tune the system and optimize its operation. The ratio pot (connected between pins J and K) allows adjustment of the ratio of the input voltages from those terminals. When the module is used as a velocity controller, the command signal connects to pin J, and the tach generator feedback connects to pin K. Because the tach generator voltage capability may be low compared with the command voltage, the ratio adjustment has the same effect as amplifying the feedback signal.

GAIN 2 pot adjusts the gain of the voltage amplifier stage. Full counterclockwise rotation gives minimum gain. Maximum clockwise rotation gives an adjustment of 13:1.

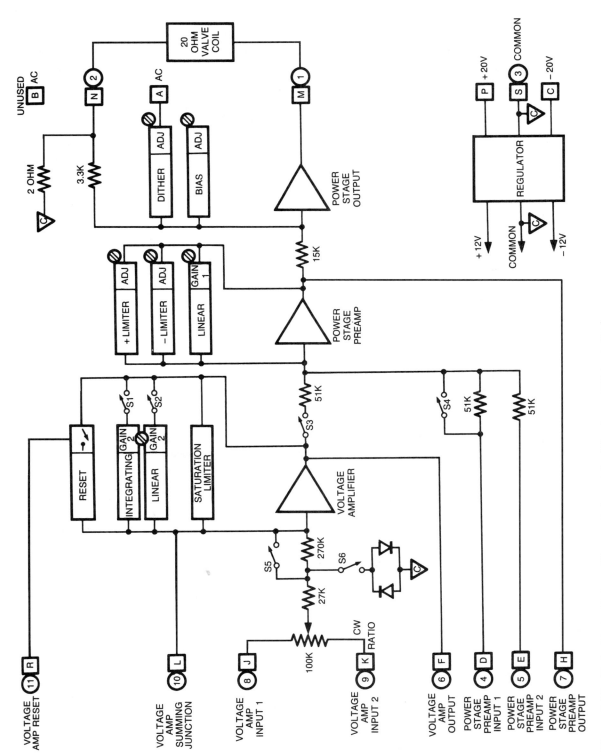

Figure 6.23 Typical industrial amplifier control module. (Courtesy of Vickers, Inc.)

150

(*Note*: This 13:1 is *not* the gain of the amplifier. It is a multiplying factor of the basic gain of the amplifier when the pot is at its minimum setting.) The GAIN 1 pot provides a 13:1 adjustment capability for the power stage preamp.

The +LIMITER and −LIMITER are used to limit the current output of the op amp to prevent damage to the torque motor. This particular amplifier is capable of driving a 20 Ω coil at 600 mA. Most torque motors are rated (maximum current) at less than 600 mA, so the output must be adjusted accordingly to prevent damage to the torque motor coils.

The BIAS pot biases the power stage output through a range from −400 mA to +400 mA. When the BIAS is used, the LIMITER adjustments are offset by the bias adjustment. If the bias was set at +200 mA, the plus limit would range from +200 mA to +600 mA, while the minus limit would range from +200 mA to −200 mA. If the module is used with bipolar valves (torque motor connected in either parallel or series arrangements), the BIAS pot should be set at the null current of the valve.

As mentioned earlier, dither is often used to improve valve performance by reducing stiction. The DITHER pot is used to adjust the *amplitude* of the dither signal from 0 to 40 mA. This is normally adjusted to 5 to 10 mA. The dither *frequency* is preset at 60 Hz.

The control module we have just discussed is normally used in conjunction with the power supply unit shown in Figure 6.24. The connector on the control module plugs into a card holder on the power supply. The external connections for typical position and velocity control circuits are shown. The numbers refer to terminals on the power supply. These correspond to the circled numbers in Figure 6.23. Other control modules may be configured differently from the one shown in Figure 6.23, but they will normally perform the same basic functions.

6.5 SERVOPUMPS AND -MOTORS

Servopumps and -motors are variable displacement units with "built-in" servocontrollers to control their displacements. Figure 6.25 shows the graphic symbols for servopumps. These are essentially miniature servo systems containing a servovalve, a hydraulic cylinder, and an internal feedback circuit.

Servopumps can be either unidirectional or bidirectional units. They are normally either bent axis or inline axial piston pumps or radial piston pumps. Their flow rates may be as high as 500 gpm, with maximum allowable pressures as high as 5000 psi, depending on pump size, physical construction, and application.

Servopumps are used almost exclusively to control the speed of large hydraulic motors by using the servo system to adjust the pump displacement in response to feedback information from the motor.

Servomotors are similar to servopumps. The servo unit is used to vary the displacement in response to command and feedback signals. The units are normally used to control motor speed, although they can also be used to control output torque.

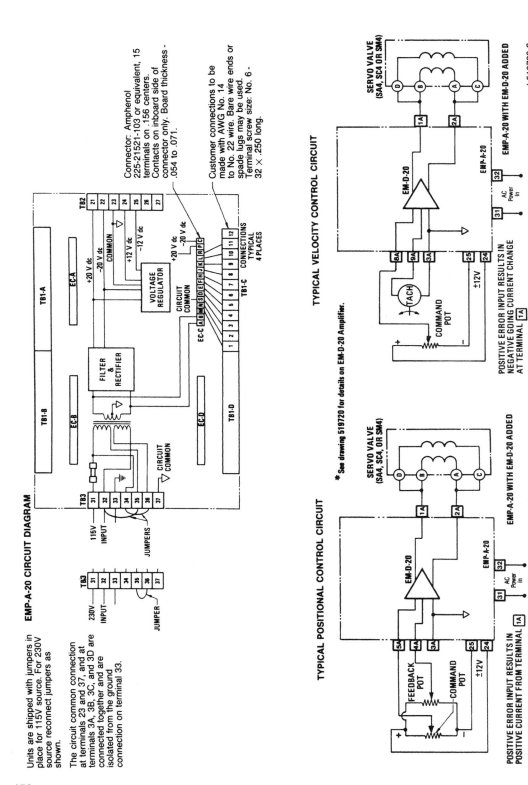

Figure 6.24 Power supply block diagram and wiring connections for the amp of Figure 6.23. (Courtesy of Vickers, Inc.)

152

Figure 6.25 Servopump symbol.

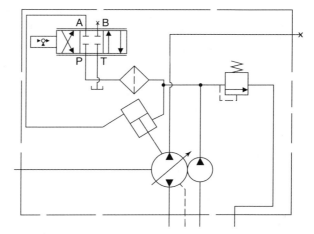

6.6   SYSTEM DESIGN CONSIDERATION

Servovalves are normally applied in the control of double-acting cylinders (both single- and double-ended) and hydraulic motors (both single and bidirectional). Because of the different areas in a single-ended cylinder, the system performance depends on the direction of travel. Although this may be acceptable in many cases, it may be more satisfactory to use a double-ended cylinder if space and other considerations allow. In addition to eliminating performance problems, the unused rod end can often be used to operate an external feedback potentiometer.

Both cylinder rod length and cylinder bore affect the repeatable error of the positioning system. The error increases with increasing length of the rod and decreases with increasing bore, which would indicate that it is preferable to use the largest bore suitable for the application. Some sources suggest that if the required stroke exceeds 10 in. (25.4 cm), a hydraulic-motor-driven rack and pinion mechanism should be considered.

Servovalves are high-pressure-drop devices and, as such, are high heat generators. This factor must be taken into consideration as follows. First, never use a valve that has a flow rating lower than that required by the system. You may be able to push more flow through it by using a higher pressure drop, but you will also generate more heat and waste more drive energy. Second, a carefully sized pressure-compensated pump is usually preferred over a fixed-displacement or even a noncompensated variable-displacement pump in a servovalve system. (The use of a pressure-compensated pump does not eliminate the need for a system relief valve.) The pump compensator should be set just slightly above the system requirement (usually 3000 psi (20 MPa)). The relief valve should be set about 10% above that. Finally, the heat generation and dissipation rates should be analyzed carefully. If the heat balance is marginal, an appropriately sized heat exchanger with automatically actuated valves should be installed.

The natural frequency of the system depends on several factors, one of which is the volume of oil under compression. In addition to properly sizing actuators to minimize the volume of oil they contain, line sizes and lengths should also be kept to a minimum. This means that the valve should be mounted as close to the actuator as possible. Many manufacturers produce cylinders and hydraulic motors with a valve mounting surface machined directly on the unit. Most cylinders of this type include internal feedback devices, some of which are very sophisticated: magnetic followers, linear resistive transducers, ultrasonic sensors, and the like. Natural frequency will be discussed in Chapter 7, and sensors will be presented in Chapter 8.

Figure 6.26 shows a very "high tech" unit. The control and valve-driving electronics, along with the central processing unit, the closed-loop control algorithm, and all the necessary feedback conditioning are located in the cover of the valve, which is mounted directly on the cylinder. It is used to control high-performance actuators where precise control of speed, acceleration, deceleration, and position are required. The valve utilizes digital closed-loop control. It can operate directly from the output of a programmable controller, personal computer, or any other device with a programmable serial port. Up to 16 of these valves can be operated on the same serial bus. Switches on the cover of each valve are set to designate the address of that particular valve.

In addition to the lines' being kept short, the line material should be rigid enough to prevent significant enlargement under high pressure. This enlargement, and subsequent relaxation, adversely affects the compliance of the system, which, in turn, reduces the frequency response of the system. Unless absolutely necessary, hoses should never be used in a servo system.

Servovalves, as a group, are far more susceptible to contamination problems than any other valve type. The small radial clearances of these valves automatically increase the probability of jamming by microscopic solid particles. If jamming does not occur, particles trapped in the clearance can significantly increase friction and cause the valve action to be sluggish or jerky.

Of more significance than the likelihood of jamming is the wear caused by contaminants. There are two critical aspects of this wear. The first is the valve leakage, which is usually low because of the radial clearances. Wear of the valve bore increases these clearances and, subsequently, the valve leakage.

An even more critical area for wear is the spool lands. Earlier discussions revealed that these are high-precision elements. If the lands are worn even slightly by the flow of solid particles across them, the performance of the valve begins to deteriorate very rapidly. For example, if the edges of the lands on a line-to-line spool begin to wear, the spool very quickly becomes an underlapped unit. Linearity, repeatability, accuracy, stability, and possibly even controllability are soon degraded to the point of unacceptability.

Figure 6.27 shows precisely such a case. The first photograph is an enlargement of an unused spool showing some minor machining marks but a well-defined edge. The second photograph shows an identical spool after it had been operated for less than 10 min in a highly contaminated system. The result of this wear was both costly and embarrassing, because it resulted in the failure of a multimillion dollar piece of equipment that was being demonstrated to a large group of government officials.

Such damage can be prevented by a carefully planned and executed contamination control program. The first step is to specify very high efficiency filters for the system, or at least the portion of the circuit containing the servovalve. Although many people recommend the use of 3-, 5-, or 10-μm filters, these designations actually have very little meaning. Filters should be specified according to their Beta ratings based on ISO 4572(2).

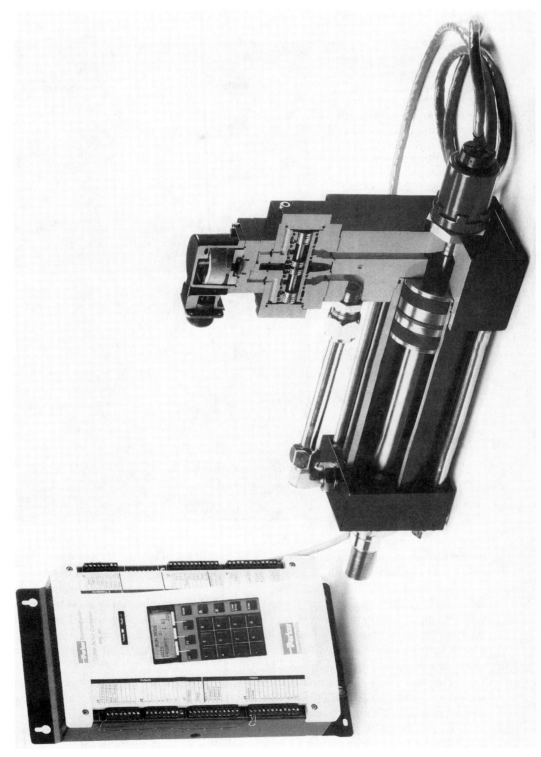

Figure 6.26 Digital servovalve mounted directly on a cylinder with an internal feedback loop. (Courtesy of Parker Hannifin Corporation.)

Figure 6.27 Effects of operating a servovalve in contaminated fluid. The top photo shows a new valve spool with the sharp edge and machining marks clearly visible. The bottom photo shows that the edge has been severely damaged by contamination. Photomicrograph 500×. (Courtesy of Thermal Control, Ltd.)

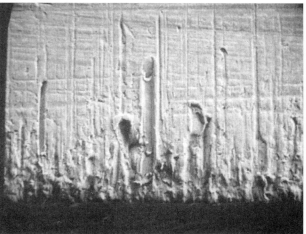

Rather than specifying the filter to be used, most manufacturers specify the fluid cleanliness level in accordance with ISO 4406(3) and leave the choice of filters to the user.

The second step in the contamination control program is properly commissioning the system before it is put into operation. This involves carefully cleaning the system reservoir, and piping, filling the system with *clean* oil (keep in mind that *new* oil is not necessarily clean), and thoroughly flushing the system at high flow and low pressure through a bank of high-Beta cleanup filters before the servovalve is installed. Special flushing blocks are available for this purpose.

The third phase of the contamination control program is a good fluid sampling and analysis program that will allow you to keep track of the fluid contamination levels and take appropriate action to prevent servovalve problems.

Finally, consider very carefully whether a servovalve is appropriate for the application. There are many applications in which an EHPV or even a basic solenoid would be acceptable. A servovalve should never be chosen simply because it seems like a good idea or because everyone else is using them in similar applications. Remember that simplicity is often preferred over sophistication, especially when the development of current technology usually outpaces the ability of the layperson to understand and use that technology properly.

6.7   SUMMARY

Servovalves offer very accurate and repeatable controls for fluid power systems because they utilize feedback information to continuously monitor and correct the output variable. Operational amplifiers are used to control the torque motor, which, in turn, controls the opening of the servovalve and the flow rate through it.

Servovalves are very sensitive to contaminants in the fluid, so it is extremely critical that the fluid be kept clean. Because servovalves are designed to operate at a high pressure drop, servo systems tend to generate a great deal of heat, which must be dissipated if fluid and system degradation is to be avoided.

In this chapter we mentioned the use of transducers, but we did not discuss them in any detail. Chapter 8 is devoted to sensors and transducers of all types. In the next chapter we will look at some of the basic system parameters that must be considered in the control and application of both proportional valves and servovalves.

REFERENCES

International Organization for Standardization. ISO 4572. 1972. Geneva.

International Organization for Standardization. ISO 4406. 1992. Geneva.

Maskrey, R. H., and Thayer, W. J. 1978. A Brief History of Electrohydraulic Mechanisms. *ASME Journal of Dynamic Systems Measurement and Control* (June) (Reprinted as *Technical Bulletin 141* by Moog Inc. Controls Division, East Aurora, NY 14052).

REVIEW PROBLEMS

General

1. Define a servovalve.
2. How do servovalves differ from proportional control valves?
3. Explain the operation of a torque motor.
4. Explain the concept of a transistor.
5. Define underlap, overlap, and line-to-line in the context of servovalve spools.
6. Define deadband.
7. Define threshold.
8. Define hysteresis.
9. List and define the types of hydraulic amplifiers (servovalve first stages).

10. Servovalves are usually rated at what pressure drop?
11. Define gain.
12. What type circuit (transfer function) is usually used for position control? Speed control?
13. A torque motor is connected in a push-pull configuration. The coils are rated at 300 mA and have a resistance of 25 ohms. Find:

    a. The voltage of each coil when the armature is centered.
    b. The maximum value of ΔI.
    c. The maximum control power for the torque motor.

14. Repeat Problem 13 for a parallel connection.
15. The valve of Problem 13 has a flow rate of 30 gpm at 3000 psi. What is its power gain?
16. A servovalve is rated at 30 gpm at 1000 psi differential. It is operated in a 2500 psi system. At the optimum power Δp, what will be its adjusted flow rate?
17. The valve of Problem 16 is used with a fluid that has a C_p of 0.5 Btu/(lb·°F) and a specific weight of 56.2 lb/ft^3. Find the heat generation rate, the horsepower loss, and the temperature rise across the valve.

CHAPTER 7

Electrohydraulic Control System Concepts

OBJECTIVES

When you have completed this chapter, you will be able to:

- Discuss Bode plots, phase lag, and amplitude response.
- Calculate the natural (resonant) frequencies for hydraulic motor and cylinder systems as well as other system parameters such as pressures and flow rates.
- Calculate the various system gains.
- Reduce complex system gains to a single expression and calculate the overall system gain.
- Determine various system parameters such as stiffness, tracking error, repeatable error, and maximum gain.
- Explain the operational differences between position control and velocity control systems.

7.1   INTRODUCTION

In Chapter 4 we looked at solenoid valves and the control systems that could be used to operate them. In Chapters 5 and 6, we discussed the two more sophisticated valves—proportional and servo—and looked briefly at their control circuitry. Now we need to look in more detail at some of the more sophisticated aspects of controlling these last two types of valves when they are used in practical systems. These concepts include the natural frequency of the system, frequency and amplitude response, and some basic considerations in feedback and control.

7.2   RAPID CYCLING

Many applications require a rapid cycling of the system. One example is a fatigue testing machine in which a hydraulic cylinder is used to produce a cyclic stress on a fatigue specimen. The objective of the test is to cyclically load the specimen to a specified test condition, then relax it to a zero load condition at the highest possible frequency. This is basically an on-off operation that can be accomplished with a standard solenoid valve; however, a higher cycling frequency and better control of the rate at which the loading is increased and decreased can be achieved using either a proportional or a servovalve. Let's see how these two valves can be used for this task.

We will assume here that the switching circuitry is not a problem. There are electrical switching devices available that can switch the electrical power on and off far faster than would be required in such a test. Therefore, our first concern deals with the frequency response of our valve. To determine that, we turn to the valve manufacturer's data sheets and examine the frequency response curves, which are usually presented in the form of a Bode plot, as shown in Figure 7.1. This Bode plot is for a typical proportional control valve.

The Bode plot presents two significant bits of information about a particular valve—its phase lag and its amplitude response—through a range of input signal frequencies. We need to understand the significance of each of these parameters. Initially, we are concerned with how rapidly we can cycle the valve, so let's look at this phase lag concept first.

7.2.1 Phase Lag

Intuitively, if not scientifically, we know that electrical responses are far faster than mechanical responses, because mass, inertia, friction, drag, external forces, and other factors

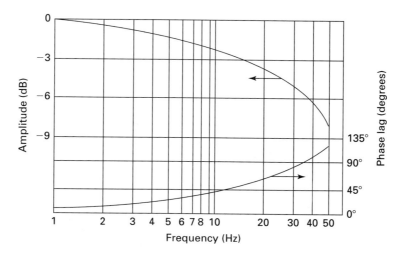

Figure 7.1 Typical Bode plot for proportional valves.

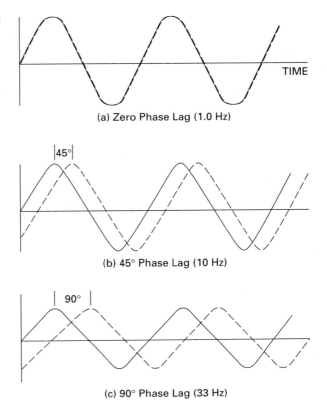

Figure 7.2 Typical phase lag in response to a sinusoidal input.

(a) Zero Phase Lag (1.0 Hz)

(b) 45° Phase Lag (10 Hz)

(c) 90° Phase Lag (33 Hz)

impede the motion of mechanical devices. Thus, it is logical to expect that at some point it is likely that electrical command signals will be supplied at a greater frequency than the mechanical device can follow. The difference is termed the *phase lag*. This concept is illustrated in Figure 7.2, which shows a sinusoidal input.

In Figure 7.2(a) the mechanical device (the valve spool in our case) is able to follow the input exactly for direction of motion. This is a zero phase lag situation. (It probably never happens, but at very low cycle rates it can come close.) Zero phase lag occurs at something less than 1 Hz on the example Bode plot of Figure 7.1.

If we increase the cycle rate to, say, 10 Hz, as shown in Figure 7.2(b), the spool begins to lag the cyclic command. Here, the spool movement lags the commanded position by an eighth of a cycle, or 45°. Thus, we say that we have a 45° phase lag at 10 Hz. This situation corresponds to the phase lag curve in Figure 7.1.

If we increase the command frequency to 33 Hz, as in Figure 7.2(c), the spool movement lags the command signal by 90°—hence, a 90° phase lag. Increasing the command frequency further results in increased phase lag. When the phase lag exceeds 180°, the valve will "go unstable," because the command signal will always be in the direction opposite that of the valve travel. This could result in a rapid, uncontrolled, and usually noisy oscillation of the spool, or the spool could "go hard over," meaning that it simply goes to one end of its stroke and stays there.

By convention, proportional valves and servovalves are rated at their 90° phase lag frequency. Thus, the valve illustrated in Figure 7.1 would be rated as a 90° phase lag at 33 Hz. It may be referred to in the literature as the "f_{90}," and it is sometimes written $f_{90} = 33$.

7.2.2 Amplitude Response

Phase lag deals with the valve's ability to follow the input command for direction. It says nothing, however, about how far the spool moves in response to the input command. This is defined for us (in relative terms) by the amplitude response curve on the Bode plot of Figure 7.1.

The amplitude response curve is the ratio of the actual output amplitude to the commanded input amplitude expressed in decibels (dB). In the case of a valve spool, we can define these amplitudes in terms of spool position. Using X_a for the actual peak position and X_c for the commanded peak position, we can define the amplitude response mathematically as

$$\text{Amplitude response} = 20 \log \left| \frac{X_a}{X_c} \right| \tag{7.1}$$

The actual response ratio can be found from Equation 7.2:

$$\frac{X_a}{X_c} = 10^{(dB/20)} \tag{7.2}$$

Example 7.1: Find the amplitude response in decibels if a valve spool moves sinusoidally 0.2 in. in response to a 0.3-in. sinusoidal command input.

Solution:

$$\text{Amplitude response} = 20 \log \left| \frac{X_a}{X_c} \right|$$

$$= 20 \log \left| \frac{0.2}{0.3} \right|$$

$$\text{Amplitude response} = -3.52 \text{ dB}$$

Looking at our example Bode plot in Figure 7.1, we can evaluate the performance of our sample valve at different input frequencies. For example, at 2 Hz the valve moves through approximately 97% (−0.3 dB) of the commanded stroke before it reverses.

$$\frac{X_a}{X_c} = 10^{(dB/20)} = 10^{(-0.3/20)} = 0.966$$

At 10 Hz it goes approximately 79% (−2 dB) of the way to the commanded position. At 20 Hz, this distance is down to only 63% (−4 dB), and drops to 50% (−6 dB) at about 35 Hz.

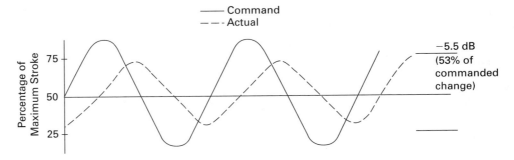

Figure 7.3 Composite frequency response to a sinusoidal input about the 50%-of-stroke position at 33 Hz.

Conventionally, valves are rated at their -3 dB amplitude response. Our valve, then, would give a -3 dB amplitude response at approximately 17 Hz. This is sometimes written as $f_{-3} = 17$.

In presenting Bode plots for valves, the manufacturer should supply information concerning the parameters used in the test procedure. For our example valve, the spool was cycled $\pm 25\%$ of its maximum stroke using 50% of the maximum stroke as its center position. This is a conventional test procedure. Based on the phase lag and amplitude response of this valve, we can plot its performance, as shown in Figure 7.3. This plot is based on the performance at 33 Hz.

7.2.3 Step Response

We have seen from the Bode plot that the valve has relatively good frequency response characteristics; however, since we want to completely relax (but not reverse) our fatigue specimen, we need to know how the valve performs about the zero position. Again, we look to the manufacturer. Somewhere in the valve data we should find some information concerning the step response of the valve. This may be tabular or graphical, but it should enable us to determine how fast the spool will move in response to a step command input. (There are many different ways to present this information, so be sure you understand what you are seeing.)

For our particular valve, the step input response is shown in Table 7.1. Here we see that in 32 ms the spool will reach 90% of its full flow (not necessarily its full stroke) in

Table 7.1

Step Input Response @ $\Delta p = 72$ psi per Metering Path	
Required Flow Step	Time to Reach 90% of Required Step
0 to 100%	32 ms
100% to 0	30 ms
+90% to −90%	50 ms

response to a step input commanding 100% flow. Notice here that the *commanded* flow is step function to 100%, whereas the time is based on an *actual* flow of 90% of the commanded flow. The valve will close in about 30 ms after the command signal is removed. This means that one cycle (zero to maximum to zero) will actually require something in excess of 62 ms. That equates to around 16 Hz.

This cycle time can be reduced if we do not require that the valve go full open on each cycle (and this may well be the case if our cylinder stroke is small). If the valve has an overlapped spool, however, a reduction in the input command amplitude will not result in a linearly proportional decrease in the cycle time. The time required to travel through the overlap deadband will remain constant, regardless of the command increment, but we will probably be able to operate the valve in excess of 16 Hz.

7.2.4 Circuit

Assuming that we can satisfy the frequency requirements, a practical circuit for a proportional valve in this application is shown in Figure 7.4. The float center of the valve should allow a complete relaxation of the stress prior to the next application. Notice that only one solenoid is connected, since load reversal is not required.

7.2.5 Natural Frequency (Cylinder Circuit)

We have seen that electronically controlled valves are capable of very high frequency oscillations. Unfortunately, just as the valves themselves are incapable of responding to the maximum frequency of the *electronic* control circuitry, the actual system is incapable of responding to the *valve's* maximum frequency capability. Let's see how this limitation affects hydraulic cylinders and motors.

All hydromechanical systems can be represented dynamically as a spring-mass system. Therefore, all such systems have a *natural* (or *resonant*) *frequency*. Any attempt to

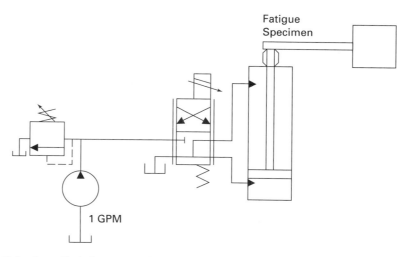

Figure 7.4   A cyclic fatigue test circuit using a proportional valve.

operate the system at any frequency higher than this natural frequency will result in erratic and unstable behavior. Because of the masses involved, the system's natural frequency will probably be much lower than the maximum operating frequency of the valve spool. Therefore, in any cyclic application, it is necessary to determine the system's natural frequency before the design can be finalized. The equation for calculating the natural frequency of a hydraulic cylinder is shown next.

$$\omega_0 = \sqrt{\frac{E}{M}\left[\frac{A_P^2}{V_P} + \frac{A_N^2}{V_R}\right]} \tag{7.3}$$

where ω_0 = natural frequency (s^{-1} or rad/s)
   E = fluid bulk modulus
   M = mass of the load (wt/g)
   A_p = piston area
   A_N = annular area (piston area − rod area)
   V_p = total volume of oil between the valve and the piston face
   V_N = total volume of oil between the annular face of the piston and the valve

Notice here the introduction of the *bulk modulus* of the hydraulic fluid. This term refers to the fact that liquids are, indeed, compressible. In the system applications we have studied so far, we have considered the fluid in the system to be incompressible, but in high-frequency cyclic applications, this is no longer permissible. The bulk modulus (which is actually the reciprocal of liquid compressibility) is calculated from Equation 7.4:

$$E = \left|\frac{-\Delta p}{\Delta V/V}\right| \tag{7.4}$$

where E = bulk modulus
   Δp = pressure change
   ΔV = volume change due to pressure application (always negative)
   V = original volume

The units of bulk modulus are the same as those of pressure (psi or MPa). The bulk moduli for petroleum-based hydraulic fluids are usually around 2×10^5 psi, while E for water is approximately 3×10^5 psi.

The values of V_p and V_N in Equation 7.3 depend on the position of the piston as it moves through its stroke. Thus, we see that the natural frequency of the system changes as the piston moves. Figure 7.5 shows that ω_0 is a maximum when the piston is fully retracted, decreases to a minimum at some point in the stroke, and then increases as the piston approaches the end of the stroke. Although the frequency can be calculated for any piston position, we are normally concerned only with the minimum value. The position at which this occurs is termed the *critical position* and can be calculated from

$$X_{\text{crit}} = \frac{(A_N S/\sqrt{A_N^3}) + (V_{LR}/\sqrt{A_N^3}) - (V_{LP}/\sqrt{A_P^3})}{(1/\sqrt{A_N}) + (1/\sqrt{A_P})} \tag{7.5}$$

where X_{crit} = critical position measured from the fully retracted position
   V_{LR} = volume of oil in the line on the rod end (cylinder to valve)
   V_{LP} = volume of oil in the line on the blind end (valve to cylinder)
   S = total cylinder stroke

Figure 7.5 Typical natural frequency versus stroke curve for a single-ended cylinder.

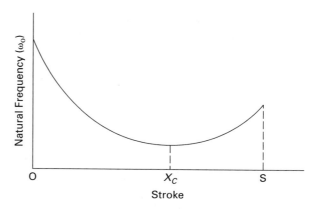

Example 7.2: The circuit in Figure 7.6 is used to oscillate the 1000-lb (4445-N) resistance as shown. The cylinder has a 1.5-in. (3.8-cm) bore and a 1.0-in. (2.54-cm) rod. It is required to move through its 24-in. (61-cm) stroke in 0.7 s. The lines between the valve and the cylinder are each 14 in. (35.6 cm) long and have a 0.76-in. (1.93 cm) ID. The fluid has a bulk modulus of 2×10^5 psi (1.4 MPa). Determine the minimum natural frequency of the system.

Solution: The minimum natural frequency will occur at the critical position. Using Equation 7.5, we obtain

$$X_{\text{crit}} = \frac{(A_N \, S/\sqrt{A_N^3}) + (V_{LR}/\sqrt{A_N^3}) - (V_{LP}/\sqrt{A_P^3})}{(1/\sqrt{A_N}) + (1/\sqrt{A_P})}$$

From the problem parameters,

$$A_P = 1.767 \text{ in.}^2$$

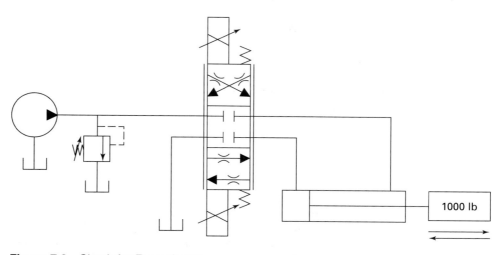

Figure 7.6 Circuit for Example 7.2.

$$A_N = A_P - A_R = (1.767 - 0.785) = 0.982 \text{ in.}^2$$

$$V_{LR} = V_{LP} = L \times A = 14 \text{ in.} \times 0.45 \text{ in.}^2 = 6.35 \text{ in.}^3$$

$$X_{\text{crit}} = \frac{\dfrac{0.982 \text{ in.}^2 \times 24 \text{ in.}}{\sqrt{(0.982 \text{ in.}^2)^3}} + \dfrac{6.35 \text{ in.}^3}{\sqrt{(0.982 \text{ in.}^2)^3}} - \dfrac{6.35 \text{ in.}^3}{\sqrt{(1.767 \text{ in.}^2)^3}}}{\dfrac{1}{\sqrt{0.982 \text{ in.}^2}} + \dfrac{1}{\sqrt{1.767 \text{ in.}^2}}}$$

$$= \frac{24.22 + 6.53 - 2.70}{1.009 + 0.7523} = 15.93 \text{ in. } (40.5 \text{ cm})$$

This means that the minimum natural frequency occurs when the piston is 15.93 in. from the fully retracted position. Using this value, we now calculate the parameters needed for Equation 7.3.

$$V_P = V_{LP} + A_P X_{\text{crit}}$$

$$= 6.35 \text{ in.}^3 + (1.767 \text{ in.}^2 \times 15.93 \text{ in.}) = 34.49 \text{ in.}^3$$

$$V_N = V_{LR} + A_N (S - X_{\text{crit}})$$

$$= 6.35 \text{ in.}^3 + 0.982 \text{ in.}^2 \times (24 - 15.93)\text{in.}^2 = 14.28 \text{ in.}^3$$

$$M = \frac{W}{g} = \frac{1000 \text{ lb}}{386 \text{ in./s}^2} = 2.59 \frac{\text{lb·s}^2}{\text{in.}}$$

Therefore,

$$\omega_0 = \sqrt{\frac{E}{M} \left[\frac{A_P^2}{V_P} + \frac{A_N^2}{V_R} \right]}$$

$$= \sqrt{\frac{2 \times 10^5 \text{ lb/in.}^2}{2.59 \text{ lb·s}^2/\text{in.}} \left[\frac{(1.767 \text{ in.}^2)^2}{34.49 \text{ in.}^2} + \frac{(0.982 \text{ in.}^2)^2}{14.28 \text{ in.}^3} \right]}$$

$$\omega_0 = 110.5 \text{ s}^{-1} = 110.5 \text{ rad/s}$$

We divide this result by 2π to get the natural frequency in cycles per second (Hz):

So
$$f_N = \frac{\omega_0}{2\pi} \tag{7.6}$$

and
$$f_N = \frac{110.5}{2\pi} = 17.58 \text{ Hz}$$

Although it appears that we have found the solution, we really have not, because we need to consider another parameter—termed *compliance*. This factor addresses the mechanical "softness" of the system. Because other components in the system are likely to be mechanically compliant (bending, stretching, and expanding piping due to hoop stresses, etc.) the usable natural frequency of the system will be reduced. The value of compliance seldom is less than 1.5, but it may exceed 10 for booms and other cantilevered loads. A commonly used value for estimating purposes when the actual value is unknown is 3. To estimate the usable natural frequency, we use

$$f_{\text{usable}} = \frac{f_N}{C} \tag{7.7}$$

where C = compliance

Since we do not know the system compliance, we will assume that $C = 3$.

Thus $f_{\text{usable}} = \frac{17.58}{3} = 5.86 \text{ Hz}$

or $\omega_{\text{usable}} = 36.8 \text{ s}^{-1}$

If we attempted to cycle the system any faster than this, we could encounter some stability problems due to limitations of the hydromechanical system, not the valve or its controller.

How could we increase the natural frequency of this system? It is really rather simple. If we look at the parameters in Equations 7.3 and 7.5, we see that the volume of oil in the lines between the valve and the cylinder is included in both equations. If we can reduce this volume of oil, we can increase the natural frequency. To accomplish this, many manufacturers are designing cylinders and motors so that the valves can be mounted directly on them, thus eliminating the connecting lines. This configuration is often termed *close coupling*.

Another design factor we want to consider is determining the flow rate required to provide the required cycle rate. To do this we first calculate the time required for the cylinder to accelerate to its maximum velocity. Because the system is essentially a spring-mass system, we begin with the equation that relates time and velocity in such a system:

$$v_{\text{act}} = v_{\text{max}} (1 - e^{-t/\tau}) \tag{7.8}$$

where v_{act} = actual velocity at any time t
           v_{max} = maximum or required velocity
              e = 2.72
              t = elapsed time
              τ = time constant for the system

For this type system we can calculate the time constant from

$$\tau = \frac{1}{\omega_{\text{crit}}} \tag{7.9}$$

where $\omega_{\text{crit}} = \frac{\omega_{\text{usable}}}{3} \tag{7.10}$

For now we disregard the actual value of t and consider only the ratio t/τ. If $t = \tau$, then $t/\tau = 1$. In this case, Equation 7.8 gives us

$$v_{\text{act}} = v_{\text{max}} (1 - e^{-t/\tau}) = 0.63 \, v_{\text{max}}$$

This means that in one time constant (τ) the cylinder will accelerate to 63% of its maximum velocity. For $t = 2\tau$, the actual velocity will be 86% of v_{max}. Ratios of 3, 4, and 5 give 95%, 96.8%, and 99.3%, respectively. Finally, when the elapsed time is six times the system time constant, the velocity will reach 99.75% of the maximum velocity. We base our calculations for acceleration on this factor of 6t. Thus, for our system, we find the total acceleration time as follows:

$$\omega_{crit} = \frac{\omega_{usable}}{3} = \frac{36.8 \text{ s}^{-1}}{3} = 12.3 \text{ s}^{-1}$$

$$\tau = \frac{1}{\omega_{crit}} = 0.082 \text{ s}$$

so that the time to a stabilized maximum velocity is

$$t_{stab} = 6\tau = (6)(0.082) = 0.492 \text{ s}$$

We can now find the maximum required velocity of the piston during extension:

$$v_{max} = \frac{S}{T_t - t_{stab}} \qquad (7.11)$$

where T_t = total time allowed to travel full stroke

$$v_{max} = \frac{24 \text{ in.}}{(0.7 - 0.492)\text{s}} = 115.4 \text{ in./s}$$

Now we can find the flow rate required to give us that velocity from

$$Q = v_{max}A_P \qquad (7.12)$$

so $$Q = \left(\frac{115.4 \text{ in.}}{s}\right)(1.767 \text{ in.}^2)\left(\frac{60 \text{ s}}{\text{min}}\right)\left(\frac{\text{gal}}{231 \text{ in.}^3}\right) = 52.96 \text{ gpm}$$

We can also calculate system pressures based on these performance parameters. First, we need the acceleration rate:

$$a = \frac{\Delta v}{t_{stab}} \qquad (7.13)$$

In this example

$$a = \frac{115.4 \text{ in./s}}{0.492 \text{ s}} = 234.6 \text{ in./s}^2 = 19.5 \text{ ft/s}^2$$

Next, we need the force required to produce this acceleration (we will neglect seal friction for now). For horizontal motion

$$F_a = Ma \qquad (7.14)$$

In this example, the force required for acceleration is

$$F_a = \frac{1000 \text{ lb} \times 19.5 \text{ ft/s}^2}{32.2 \text{ ft/s}^2} = 607 \text{ lb}$$

Seal friction can be important, so we should include it in our calculation:

$$F_t = F_a + Wf_s$$

where f_s = seal coefficient of friction (assume 0.3)
         W = weight of the load

This gives

$$F_t = 607 + (1000)(0.3) = 907 \text{ lb}$$

The pressure required for acceleration during extension is

$$p = \frac{F_t}{A_P} = \frac{2045 \text{ lb}}{1.767 \text{ in.}^2} = 513 \text{ psi}$$

To find the pressure required for acceleration during retraction, we would use similar calculations using the annular area.

If we assume that deceleration at each end of the stroke occurs at the same rate, then these same pressures will result at the ends of the stroke. (Only the pressure required to overcome seal resistance will be needed once the piston reaches its maximum velocity.) If, however, the DCV closes before the piston reaches the end of its stroke, and if the closing time of the valve is less than the deceleration time of the piston, much higher pressures can be generated by pressure intensification.

Consider that in our example system the valve closed in 0.016 s. The acceleration rate (really, deceleration rate) would then be

$$a = \frac{\Delta v_M}{T_S} \qquad (7.15)$$

where T_S is the spool closing time. Then

$$a = \frac{115.4 \text{ in./s}}{0.016 \text{ s}} = 7212.5 \text{ in./s}^2 = 601.0 \text{ ft/s}^2$$

and

$$F = Ma = \frac{1000 \text{ lb} \times 601.0 \text{ ft/s}^2}{32.2 \text{ ft/s}^2} = 18{,}665 \text{ lb}$$

so that

$$p = \frac{F}{A_N} = \frac{18{,}665 \text{ lb}}{0.982 \text{ in.}^2} = 19{,}007 \text{ psi}$$

in the rod end of the cylinder and the pipe between the cylinder and the valve. This might cause problems in a system designed for 3000 psi.

The ramping function on an EHPV or the gain setting on a servovalve could be used to alleviate this problem. For instance, if we wanted to hold the maximum pressure to 3000 psi, we could solve Equation 7.15 for T_S to determine the minimum valve closing time we could tolerate. Thus,

$$T_S = \frac{v_{max}}{a} \qquad (7.16)$$

Working backwards, we get

$$F_{max} = p \times A_N = (3000 \text{ lb/in.}^2)\,(0.982 \text{ in.}^2) = 2946 \text{ lb}$$

$$a = \frac{F_{max}}{M} = \frac{2946 \text{ lb}}{1000 \text{ lb}/32.2 \text{ ft/s}^2} = 94.86 \text{ ft/s}^2$$

$$T_S = \frac{v_{max}}{a} = \frac{(115.4 \text{ in./s})(\text{ft}/12 \text{ in.})}{94.86 \text{ ft/s}^2} = 0.1014 \text{ s}$$

Thus, if we adjust the ramp to give us a shifting time of about 0.1 s, we will not exceed 3000 psi.

7.2.6 Natural Frequency (Hydraulic Motor Circuit)

The natural frequency for a hydraulic motor circuit is found in much the same way as for a cylinder. Here,

$$\omega_0 = \sqrt{\frac{2E \times (V_M/\pi)^2}{V_T I_T}} \qquad (7.17)$$

where V_M = hydraulic motor displacement
V_T = effective trapped volume of fluid
I_T = total reflected inertia

The effective trapped volume is found from Equation 7.18:

$$V_T = V_M + V_L \qquad (7.18)$$

where V_L = volume of fluid in both the pressure and return lines between the valve and motor

The total reflected inertia from the load is

$$I_T = Mr^2 R^2 \qquad (7.19)$$

where r = radius of gyration (or drive gear radius if used)
R = gear ratio (if used) = $\dfrac{\text{number of teeth on input gear}}{\text{number of teeth on output gear}}$

Example 7.3: Consider the circuit shown in Figure 7.7. The hydraulic motor operates at 60 rpm to turn a 1000-lb (4445-N) load with a radius of gyration of 1 ft (0.305 m). The motor has a 60 in.^{3}/rev (983 cm^3/rev) displacement. The total length of the lines is 12 ft (3.66 m) with a 0.56-in. (1.4-cm) ID. The bulk modulus of the oil is 2×10^5 psi (1.4 MPa), and the maximum system pressure is 3000 psi (21 kPa). Find the natural frequency of the system.

Solution: From Equation 7.18, the trapped fluid volume is

$$V_T = V_M + V_L$$

$$V_L = AL = (0.246 \text{ in.}^2)(12 \text{ ft})(12 \text{ in./ft}) = 35.42 \text{ in.}^3$$

$$V_T = 60 \text{ in.}^3 + 35.42 \text{ in.}^3 = 95.42 \text{ in.}^3$$

From Equation 7.19, the reflected inertia is

$$I_T = Mr^2 R^2$$

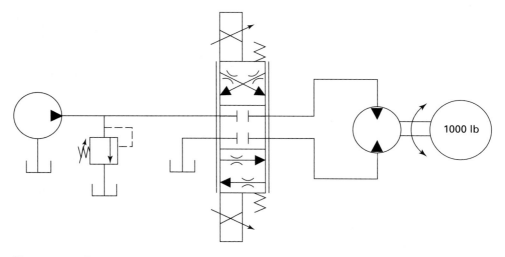

Figure 7.7 Circuit for Example 7.3.

Since no gearing is used, $R = 1$, so

$$I_T = \frac{(1000 \text{ lb})(1 \text{ ft})^2}{32.2 \text{ ft/s}^2}$$

$$= 31.06 \text{ lb·ft·s}^2$$

$$= 372.67 \text{ lb·in.·s}^2$$

Therefore, from Equation 7.17,

$$\omega_0 = \sqrt{\frac{2E \times (V_M/\pi)^2}{V_T I_T}}$$

$$= \sqrt{\frac{2(2 \times 10^5 \text{ lb/in.}^2)(60 \text{ in.}^3/\pi)^2}{(95.42 \text{ in.}^3)(372.67 \text{ lb·in.·s}^2)}}$$

$$\omega_0 = 64.04 \text{ s}^{-1} = 64.04 \text{ rad/s}$$

Again, we need to consider system stiffness. Using a compliance factor of 3, we find the usable frequency to be

$$\omega_{\text{usable}} = \frac{\omega_0}{C} = \frac{64.04}{3} = 21.3 \text{ rad/s}$$

or

$$f_{\text{usable}} = \frac{\omega_{\text{crit}}}{2\pi} = 3.4 \text{ Hz}$$

The critical frequency is one-third of the usable frequency, or 7.1 rad/s.

As with hydraulic cylinders, there are numerous other design parameters that we can calculate for hydraulic motors. For example, we can use Equations 7.9 and 7.11 to find the acceleration time and rate, respectively.

So
$$\tau = \frac{1}{\omega_{\text{crit}}} = \frac{1}{7.1} = 0.14 \text{ s}$$

and
$$a = \frac{v_M}{t_{\text{stab}}}$$

The v_M we use here is a rotary velocity and must be converted to rad/s.

$$v_M = \frac{\text{rpm}}{60} \times 2\pi = \frac{60}{60} \times 2\pi = 2\pi \text{ rad/s}$$

Again,
$$t_{\text{stab}} = 6\tau = (6)(0.14) = 0.85 \text{ s}$$

Thus,
$$a = \frac{2\pi \text{ rad/s}}{0.85 \text{ s}} = 7.39 \text{ rad/s}^2$$

We find the starting torque required from

$$T = I_T a \tag{7.20}$$
$$= (31.06 \text{ lb·ft·s}^2)(7.39 \text{ rad/s}^2)$$
$$= 229.6 \text{ lb·ft}$$

The pressure required to produce the starting torque is (assuming a mechanical efficiency of 0.95):

$$p = \frac{2\pi T}{V_M \eta_M} \tag{7.21}$$
$$= \frac{2\pi(229.6 \text{ lb·ft})(12 \text{ in./ft})}{(60 \text{ in.}^3)(0.95)} = 303.7 \text{ psi}$$

The flow rate required at the running speed (assuming a volumetric efficiency of 0.9) is

$$Q = \frac{V_M N}{\eta_V} \tag{7.22}$$
$$= \frac{(60 \text{ in.}^3/\text{rev})(60 \text{ rev/min})}{0.9} = 4000 \text{ in.}^3/\text{min} = 17.32 \text{ gpm}$$

The values found in these two examples could be used to determine other factors in the design process. For example, the pressures calculated have not included Bernoulli-type losses due to friction in pipelines and fittings. The calculated flow rates would be used for finding these losses. Pressure drops across the valves have not been included. These can be found from the valve manufacturer's data sheets.

Once total pressure and flow requirements are known, horsepower requirements can be calculated for each part of the machine duty cycle. From this information, the peak and rms power requirements of the system can be determined. The peak horsepower will be based on the maximum requirement for any phase of the operation per Equation 7.23:

$$\text{HHP} = \frac{p \times Q}{1714} \tag{7.23}$$

The rms horsepower uses the horsepower required during each phase of the process and duration of each phase in Equation 7.24:

$$\text{HHP}_{\text{RMS}} = \sqrt{\frac{\Sigma_1^n \, (\text{HHP}_n \times t_n)}{\Sigma_1^n t_n}} \tag{7.24}$$

These are horsepowers in the fluid system itself. The input horsepower from the prime mover is based on the overall efficiency of the hydraulic pump:

$$\text{IHP} = \frac{\text{HHP}}{\eta_O} \tag{7.25}$$

If an electric motor is used for an industrial system, the motor must be able to supply the required input horsepower during its normal operation but have enough reserve to satisfy the peak horsepower requirement. A good rule of thumb is that this peak requirement should be met at 90% of the rated voltage.

The National Electrical Manufacturers Association (NEMA) defines the overload operating capability of electric motors by the percentage of breakdown torque (BDT) at which the motor can be operated before it stalls (and ultimately fails). Using the 90% rated voltage factor, we can calculate the percent BDT using Equation 7.26:

$$\%\text{BDT} = \frac{\text{Peak-load HP}}{\text{Nameplate HP}} \times \frac{100}{(0.9)^2} \tag{7.26}$$

$$= \frac{\text{Peak-load HP}}{\text{Nameplate HP}} \times 123$$

As an added safety factor, many designers add an additional 20% to this value, to give

$$\%\text{BDT} = \left[\frac{\text{Peak-load HP}}{\text{Nameplate HP}} \times 123 \right] + 20 \tag{7.27}$$

Thus, to meet the peak system requirements, the percent BDT at which the motor is rated must exceed that calculated by Equation 7.27.

7.3   GAIN AND FEEDBACK

One of the most important concepts of a servo system is its ability to constantly monitor its output and automatically make corrections to ensure that the output remains at the commanded level. This is accomplished through the use of some type of feedback from a transducer (see Chapter 8) that monitors the output parameter. We have already looked at this concept in the context of the valve. Now let's expand the concept to include the entire system.

7.3.1 The Control Ratio Equation

One of the most important aspects of system analysis is an understanding of system gain. We have already discussed gain, but only in the limited context of the electronics. Now

Figure 7.8 A basic feedback (closed-loop) circuit.

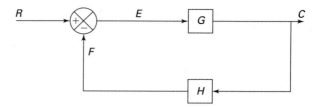

let's broaden our scope to include the entire system. For simplicity, we will consider the individual element gains to be linear, although we have already seen that this is not necessarily the case.

Figure 7.8 is a block diagram of a generic system with feedback. The *command input signal* is given the designator *R* for reference. The gain of the top leg is designated *G* and termed the *forward loop gain*. The gain of the *feedback loop* is denoted by *H*. The *controlled output* is shown as *C*.

As the system operates, a feedback signal *F* is continuously generated. Its value is based on the value of *C* and the feedback gain. Thus,

$$F = CH \tag{7.28}$$

This feedback signal is fed to the summing junction (in the op amp), where it is compared with the reference signal. The result is an error signal *E*, whose value is

$$E = R - F = R - CH \tag{7.29}$$

This results in a change in the controlled variable *C*, which then becomes

$$C = EG \tag{7.30}$$

and the process repeats itself.

By equating Equations 7.29 and 7.30 we get

$$C = (R - CH)G$$

$$= RG - CGH$$

so that $\qquad\qquad RG = C + CGH = C(1 + GH)$

And, eventually,

$$\frac{C}{R} = \frac{G}{1 + GH} \tag{7.31}$$

This equation is known as the *control ratio*, the *closed-loop gain*, or the *closed-loop transfer function* of the system. You can see that it represents the output divided by the input, which we have previously identified as gain. Therefore, the right-hand side of Equation 7.31 defines the system gain and is commonly referred to as the *closed-loop gain* of the system.

Example 7.4: In the system of Figure 7.8, suppose the forward loop contains a hydraulic motor. Let's say that the forward loop gain is 300 rpm/V and that the feedback loop consists of a tach generator that has a gain of 0.2 V/rpm. What will be the speed of the hydraulic motor for an input of 3.5 V?

Solution: From the problem statement

$$R = 3.5 \text{ V}$$
$$G = 300 \text{ rpm/V}$$
$$H = 0.2 \text{ V/rpm}$$

Solving Equation 7.31 for C gives us

$$C = \left[\frac{G}{1 + GH}\right] R$$

$$= \left[\frac{300 \text{ rpm/V}}{1 + (300 \text{ rpm/V})(0.2 \text{ V/rpm})}\right] 3.5 \text{ V}$$

$$C = 17.2 \text{ rpm}$$

This may not be the answer you expected. At first glance, we see that since the motor turns 300 rpm for every volt of input, a 3.5 V input should give 1050 rpm. This would indeed be the case if we had no feedback, because only the forward loop gain would be used; however, we do have feedback, so Equation 7.31 must be used.

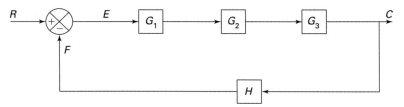

Figure 7.9 Multiple gains in the forward loop.

Now let's expand the system of Figure 7.8 to include more elements in the forward loop, as in Figure 7.9. We can still find the gain of this circuit by using Equation 7.31, but first, we need to resolve all the forward loop gains into a single value of G. The forward loop gain G is the product of all the individual element gains between the summing junction and the output. Thus,

$$G = G_1 G_2 G_3$$

so that

$$\frac{C}{R} = \frac{G}{1 + GH} = \frac{G_1 G_2 G_3}{1 + G_1 G_2 G_3 H}$$

Example 7.5: Determine the speed of the hydraulic motor in the circuit shown in Figure 7.10.

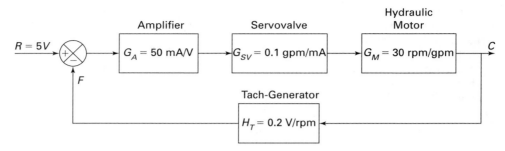

Figure 7.10   Block diagram for Example 7.5.

Solution: Resolving the forward loop gain gives

$$G = G_A G_{SV} G_M$$

$$= (50 \text{ mA/V})(0.1 \text{ gpm/mA})(30 \text{ rpm/gpm})$$

$$= 150 \text{ rpm/V}$$

Now, using Equation 7.31, we get

$$C = \left[\frac{G}{1 + GH} \right] R$$

$$= \left[\frac{150 \text{ rpm/V}}{1 + (150 \text{ rpm/V})(0.2 \text{ V/rpm})} \right] 5 \text{ V}$$

$$= 24.2 \text{ rpm}$$

7.3.2 Multiple Feedback Systems

We often find systems that have *internal* feedback loops. For instance, a servovalve may use an internal LVDT to feed back spool position. How would we handle a system such as the one shown in Figure 7.11a? Again, it is simply a matter of resolving the gains to get back to G and H in Equation 7.31.

First, we tackle the inner loop, applying Equation 7.31 as if this small loop were a complete system. We get as its equivalent gain,

$$\frac{G_2}{1 + G_2 H_1}$$

We can substitute this value for that small loop as in Figure 7.11b.

Next, we multiply the individual forward loop gains together to get a single expression:

$$G = \frac{G_1 G_2 G_3}{1 + G_2 H_1}$$

Figure 7.11 (a) Basic closed-loop circuit with internal feedback loop. (b) Basic closed-loop circuit with internal feedback loop resolved. (c) Basic closed-loop circuit with entire forward loop resolved.

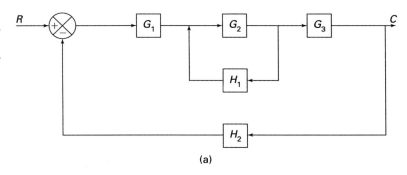

(a)

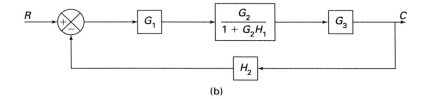

(b)

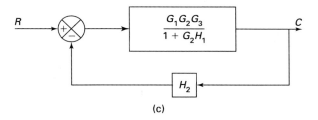

(c)

Now we are back to the basic system of Figure 7.9. Applying Equation 7.31 again, we get

$$\frac{C}{R} = \frac{\dfrac{G_1 G_2 G_3}{1 + G_2 H_1}}{1 + \dfrac{G_1 G_2 G_3 H_2}{1 + G_2 H_1}}$$

which eventually reduces to

$$\frac{C}{R} = \frac{G_1 G_2 G_3}{1 + G_2 H_1 + G_1 G_2 G_3 H_2}$$

Although this expression looks rather simple, bear in mind that these Gs and Hs may very well represent transfer functions corresponding to first- and second-order differential equations. We will not get into that aspect, but if you find yourself needing to model such a system, a good text in control theory would be very useful.

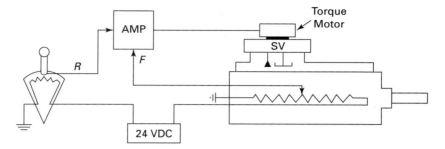

Figure 7.12 Servo-controlled positioning system for the example of Section 7.3.3.

7.3.3 A System Example

At this point it would seem that we are ready to look at a system analysis exercise. Some of the exercise will use concepts we have already covered, but there will be some new material, also. Follow the exercise carefully.

Figure 7.12 is a servo-controlled positioning system. The input command comes from a joystick controller that has a 90° throw. The input to the controller is 24 VDC. The cylinder has a 4-in.^{2} (26-cm^2) piston and a 2-in.^{2} (12.9-cm^2) rod with a 6-in. (15.24-cm) stroke. The position transducer is a linear resistive transducer (LRT) or potentiometer that uses a 24 VDC power supply. The operational amplifier has an output of 500 mA for each input volt; however, the torque motor on the servovalve is rated at 200 mA, at which point the valve is fully open. (This is the *saturation* current of the torque motor.) Thus, the current limiter of the op amp is set to a maximum output of 200 mA, regardless of the input, to prevent burning out the torque motor. The hydraulic system working pressure is 3000 psi (20 MPa) with a flow rate of 5 gpm (18.9 lpm). At a torque motor input of 10 mA, the pressure differential across the valve is 1000 psi (6.9 MPa).

This seems to be a real jumble of information. The first thing we need to do is sort out some of it into more useful groupings. For example, we can find the gain of the joystick controller by assuming that the full 24 V is linearly distributed throughout its stroke. Thus,

$$G_{js} = \frac{24 \text{ V}}{90°} = 0.267 \text{ V/°}$$

Notice that we found this gain by using the maximum voltage divided by the maximum rotation. Also, we assumed that the gain is linear throughout the full throw. We can find the LRT gain in the same manner. Since this is a feedback, we write

$$H = \frac{24 \text{ V}}{6 \text{ in.}} = 4 \text{ V/in.}$$

The gain of the operational amplifier is given as

$$G_A = 500 \text{ mA/V}$$

The valve flow gain is the flow through the valve per milliamp of current input. Here, we use the maximum values of both flow and current to get

$$G_{SV} = \left(\frac{5 \text{ gpm}}{200 \text{ mA}}\right)\left(\frac{3.85 \text{ in.}^3/\text{s}}{\text{gpm}}\right) = 0.096 \left(\frac{\text{in.}^3/\text{s}}{\text{mA}}\right)$$

which means that every milliamp of current to the torque motor will cause the valve to open far enough to allow an additional 0.096 in.^{3}/s of flow to the cylinder. The valve *pressure gain* is defined as the pressure drop across the valve at a given current input. In this case,

$$G_P = \frac{1000 \text{ psi}}{10 \text{ mA}} = 100 \frac{\text{psi}}{\text{mA}}$$

The *position gain* of the cylinder defines the distance the piston will move for each unit volume of fluid put into it. The apparent way to calculate this value would be

$$G_{\text{cyl}} = \frac{d}{V}$$

where d = distance moved
         V = volume of fluid

We can also define this position gain based on the cylinder velocity and the volume flow rate. In this case,

$$G_{\text{cyl}} = \frac{v}{Q}$$

However, we can define the flow rate as the area multiplied by the velocity:

$$Q = Av$$

in which case we see that position gain can also be calculated from

$$G_{\text{cyl}} = \frac{v}{Q} = \frac{v}{Av} = \frac{1}{A}$$

We actually have a choice of units for position gain (an unusual luxury). If we are interested in position only, the units we use are in./in.^{3} or cm/cm^3. If we are concerned with velocity, we use (in./s)/(in.^{3}/s) or (cm/s)/(cm^3/s). This latter choice can also be termed the *velocity gain* of the cylinder. Both choices are numerically equal. In this case the position gain of the cylinder

$$G_{\text{cyl}} = \frac{v}{Q} = \frac{1}{A} = 0.25 \frac{\text{in./s}}{\text{in.}^3/\text{s}} = 0.25 \frac{\text{in.}}{\text{in.}^3}$$

Now, we can draw a block diagram of the system (Figure 7.13). Notice that the gain of the cylinder is shown in in./in.^{3} because we are concerned about position rather than velocity. Likewise, we drop the time (seconds) in the flow rate. Thus, the cylinder will move 0.25 in. for every cubic inch of oil that enters it.

We can now find the control ratio for the system using Equation 7.31:

$$\frac{C}{R} = \frac{G}{1 + GH}$$

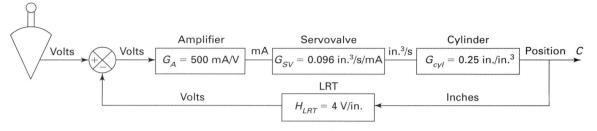

Figure 7.13 Block diagram of the system of Figure 7.12.

$$C = G_A G_{SV} G_{cyl}$$
$$= (500 \text{ mA/V})(0.096 \text{ in.}^3/\text{mA})(0.25 \text{ in./in.}^3)$$
$$= 12 \text{ in./V}$$

Notice here that the gain of the joystick is *not* included in the forward loop gain.

Therefore, $\quad \dfrac{C}{R} = \dfrac{12 \text{ in./V}}{1 + (12 \text{ in./V})(4 \text{ V/in.})} = 0.245 \text{ in./V}$

This means that for every volt of input, the cylinder will extend 0.245 in. (0.622 cm). We can now calculate the new cylinder position for any movement of the joystick. For example, if we move the joystick 1.5° from its stop, the input command R will be

$$R = (0.267 \text{ V/°})(1.5°) = 0.4 \text{ V}$$

and the new position will be

$$C = \frac{C}{R}R$$

$$C = (0.245 \text{ in./V})R = (0.245 \text{ in./V})(0.4 \text{ V}) = 0.098 \text{ in. } (0.25 \text{ cm})$$

Although the choice of 1.5° may seem odd, it was a deliberate choice, because it leads us to a more detailed analysis. Suppose you were able to move the joystick through that 1.5° instantaneously and watch the reaction of the system. What would you see?

Step by step, here is what would happen. The 1.5° movement would generate a 0.4 V command signal. Because the cylinder had not yet moved, the feedback would be zero. This would result in an error signal of 0.4 V from the summing junction to the amplifier, which would, in turn, send a [(0.4 V)(500 mA/V) =] 200 mA signal to the torque motor. Because this is the saturation current for the torque motor, it would open the valve fully and cause the cylinder to accelerate toward full speed. As soon as it started to move, however, a feedback signal would be generated, the error signal would begin to drop, and the valve would start to close, which would affect the acceleration rate. The cylinder would stop when the feedback signal matched the command signal, 0.1 in. from its starting point, which is approximately where our calculations predicted it would be.

If you had to predict the final position of the cylinder, you might be able to think of an easier method than we have just used. The objective of the entire operation is to position the cylinder at a point where the feedback signal equals the command signal. Apparently, we could easily find that position by setting these two signals equal and solving for the position. Thus,

$$\text{Input} = H \times \text{displacement}$$

or

$$\text{Displacement} = \frac{\text{input}}{H}$$

Since moving the joystick 1.5° gave a 0.4 V command signal, the cylinder would move

$$\frac{0.4 \text{ V}}{4 \text{ V/in.}} = 0.1 \text{ in. (0.254 cm)}$$

We can probably agree that this is an exact calculation, whereas our control ratio equation is just a sophisticated estimate.

Now, although the position calculated using Equation 7.31 gave fairly close results, was it due to a judicious selection of gains in the forward loop, or is the equation really that good? Let's change some values and see.

Suppose we increase the amplifier gain to 100 mA/V. Now G becomes 24 in./V, giving us a control ratio of

$$\frac{C}{R} = \frac{24 \text{ in./V}}{1 + (24 \text{ in./V})(4 \text{ V/in.})} = 0.247 \text{ in./V}$$

So our final position is

$$C = (0.247 \text{ in./V})(0.4 \text{ V}) = 0.098969 \text{ in. (0.25138 cm)}$$

which is very close to our previous result.

Let's also increase the flow gain to 0.2 in.^{3}/s/mA. Now G is 50 in./V, the control ratio is 0.248 in./V, and the calculated final position is 0.0995 in. (2.5273 cm).

If, in addition to these changes, we increase the cylinder gain to 0.5 in./in.^{3}, G becomes 100 in./V, the control ratio becomes 0.249 in./V, and the final position is estimated to be 0.09975 in. (0.25337 cm).

Thus, it becomes obvious that the gains in the forward loop have little to do with the final position of the cylinder in this type of cylinder. The real significance is that they determine how rapidly the final position is reached and how stiff the system is. As we will see a little later, these gains, especially the amplifier gain, definitely affect other aspects of the system operation, including stability, response time, and repeatable error. You might also notice that a change in the feedback gain would significantly affect the control ratio, but this is impractical in this case, because the available cylinder stroke would be affected. For example, doubling the feedback gain would cause the feedback signal to equal the maximum command signal when the cylinder had extended to only half its full stroke.

Returning to our circuit observation, let's very quickly move the joystick from 1.5° to 45°. This creates a command signal of 12 V. Since the feedback signal was at 0.4 V due to the earlier movement, the new error signal is 11.6 V. That apparently generates an amplifier output of [500 mA/V $\times$ 11.6 V] = 5800 mA. Recall, though, that the amplifier saturates at 200 mA and cannot put out a larger signal. Therefore, the valve will go full open and stay there until the error signal drops below that required to produce a 200 mA amplifier output. This occurs when the error signal voltage causes the amplifier to reach the saturation current, which is found from Equation 7.32:

$$E_{\text{sat}} = \frac{I_{\text{sat}}}{G_A} \tag{7.32}$$

In this case,

$$E_{\text{sat}} = \frac{200 \text{ mA}}{500 \text{ mA/V}} = 0.4 \text{ V}$$

The position at which this occurs measured from the final (stopping) position is found from

$$X_{\text{dec}} = \frac{E_{\text{sat}}}{H}$$

Here,

$$X_{\text{dec}} = \frac{0.4 \text{ V}}{4 \text{ V/in.}} = 0.1 \text{ in. (0.254 cm)}$$

from the final position. Thus, the cylinder travels at full speed until it is within 0.1 in. (0.254 cm) of its final position, then decelerates and stops at its final position. The final position occurs when the feedback signal equals the command signal—3 in. (7.62 cm) in this case.

The significance of the pressure gain becomes apparent when a disturbing force is considered. Suppose a force of 1000 lb (4445 N) is applied on the end of the rod, which attempts to push the cylinder back in. This will result in a cylinder pressure of [1000 lb/ 4 in.^{2} =] 250 psi (67 MPa). In order to compensate for this pressure and prevent movement of the cylinder, on offset current must be applied to the valve torque motor. The value of this offset current is found using Equation 7.33:

$$I_{\text{offset}} = \frac{P_{\text{cyl}}}{G_p} \tag{7.33}$$

For this situation, an offset current of

$$I_{\text{offset}} = \frac{250 \text{ psi}}{100 \text{ psi/mA}} = 2.5 \text{ mA}$$

would be required. To produce this offset current would require an offset error signal of

$$E_{\text{offset}} = \frac{I_{\text{offset}}}{G_a} \tag{7.34}$$

or

$$E_{\text{offset}} = \frac{2.5 \text{ mA}}{500 \text{ mA/V}} = 0.005 \text{ V}$$

The only way this error signal can be produced is to displace the cylinder, causing a displacement in the potentiometer. The amount of displacement required to produce the error signal is calculated using Equation 7.35:

$$X_{\text{offset}} = \frac{E_{\text{offset}}}{H} \tag{7.35}$$

In our example, this displacement is

$$X_{\text{offset}} = \frac{0.005 \text{ V}}{4 \text{ V/in.}} = 0.00125 \text{ in. } (0.00318 \text{ cm})$$

This value defines the *system stiffness*, that is, the maximum disturbance that would not be corrected.

7.3.4 Gain Adjustments

Now let's look at the effect of the amplifier gain on the performance of the system. Using Equations 7.32 and 7.33 we can show that if we increased the amplifier gain to 1000 mA/V, an error signal of only 0.2 V would saturate the amp and push the valve wide open. It would stay wide open until the cylinder was within 0.05 in. of its stopping point. The system stiffness would improve to 0.000625 in., which would result in a very stiff and fast-acting system.

If we reduced the amplifier gain to 250 mA/V, the stopping distance would increase to 0.2 in. and the stiffness to 0.0025 in. Too low a gain will result in a sluggish, spongy system that will tend to "hunt" for the commanded position.

It is obvious that increasing the amplifier gain can significantly improve the performance of the system. By reducing the stopping distance, the time required to move from one position to another is decreased. This improvement does not come without penalty, however. As we saw in Chapter 5, extremely high pressure can result from a rapidly closing valve, which can cause plumbing or components to rupture. Another result can be a phenomenon known as *water hammer*, which is caused by suddenly stopping a flowing liquid by rapidly closing a valve.

System stability can also be adversely affected by high amplifier gain. As the gain is increased the sensitivity of the system to disturbances becomes extremely high. The result is a tendency to overcorrect, then abruptly reverse and overcorrect in the opposite direction. An unstable system will oscillate wildly and can cause serious damage to itself and the machine on which it is installed. The maximum allowable gain of a system can be estimated mathematically if certain information about the components making up the system is known (and it usually is).

Let's look back to the circuit in Figure 7.13 and try to determine the maximum amplifier gain we could use without causing the system to go unstable. First, we need to calculate the open-loop gain. (We have looked only at the closed-loop gain to this point.) This is readily found by multiplying all the circuit gains, both forward loop and feedback.

$$\text{Open-loop gain} = G_A G_{SV} G_{\text{cyl}} H \tag{7.36}$$

This is also termed the *velocity constant* and given the symbol K_V, so we have

$$K_V = G_A G_{SV} G_{\text{cyl}} H \tag{7.37}$$

Solving for the amplifier gain gives us

$$G_A = \frac{K_V}{G_{SV}G_{cyl}H} \tag{7.38}$$

The velocity constant of the system is the reciprocal of the time constant that we found earlier in this chapter when we calculated the maximum usable frequency for a circuit, so we can also calculate K_v from

$$K_V = \frac{1}{\tau} \text{ s}^{-1} \tag{7.39}$$

We can use Equation 7.3 to calculate the natural frequency of our circuit:

$$\omega_0 = \sqrt{\frac{E}{M}\left[\frac{A_P^2}{V_P} + \frac{A_N^2}{V_R}\right]}$$

In this example, we use a bulk modulus of 2×10^5 psi (1.4 MPa) and a load weighing 3860 lb (17 kN). This gives us a mass of

$$M = \frac{wt}{g} = \frac{3860 \text{ lb}}{386 \text{ in./s}^2} = 10 \frac{\text{lb·s}^2}{\text{in.}}$$

We assume that the piston is in the critical position, at which point $V_P = 30$ in.^{3} and $V_N = 20$ in.^{3}. These are assumed values; we have not actually calculated them for this example.
From Equation 7.3,

$$\omega_0 = \sqrt{\frac{2 \times 10^5 \text{ lb/in.}^2}{10 \text{ lb·s}^2/\text{in.}}\left[\frac{16 \text{ in.}^4}{30 \text{ in.}^5} + \frac{4 \text{ in.}^4}{20 \text{ in.}^3}\right]}$$

$$\omega_0 = 121 \text{ s}^{-1}$$

Assuming a compliance of 3, we have a usable frequency of 40.37 s^{-1}, a critical frequency of 13.4 s^{-1}, and a time constant of 0.075 s. From Equation 7.39, we find that the velocity constant for this circuit is

$$K_V = \frac{1}{\tau} = \frac{1}{0.075} = 13.4 \text{ s}^{-1}$$

Now, returning to Equation 7.38, we find that the maximum value for G_A is

$$G_A = \frac{K_V}{G_{SV}G_{cyl}H}$$

$$= \frac{13.4 \text{ s}^{-1}}{\left[\dfrac{0.096 \text{ in.}^3/\text{s}}{\text{mA}}\right]\left[\dfrac{0.25 \text{ in.}}{\text{in.}^3}\right]\left[\dfrac{4 \text{ V}}{\text{in.}}\right]}$$

$$G_A = 140 \text{ mA/V}$$

This represents the theoretical maximum allowable amplifier gain. Setting the amplifier gain higher than this could result in system instability. In practice, the gain may have to be set at a lower value. In our example, the amplifier gain is 500 mA/V, so we are probably in trouble.

7.3.5 Repeatable Error

Repeatable error defines the ability of the system to return to the same position every time a specific command input is provided. The repeatable error is a function of the deadband of the valve and the electronic gain of the system, as shown in Equation 7.40:

$$R_e = \frac{\text{deadband (mA)}}{G_A\left(\frac{\text{mA}}{\text{V}}\right)H\left(\frac{\text{V}}{\text{in.}}\right)} \tag{7.40}$$

Looking again at the system of Figure 7.13, we see that the amplifier gain was 500 mA/V while the feedback gain was 4 V/in. A typical deadband is around 5 mA. Substituting these values into Equation 7.40 gives us

$$R_e = \frac{5 \text{ mA}}{\left(500\frac{\text{mA}}{\text{V}}\right)\left(4\frac{\text{V}}{\text{in.}}\right)} = 0.0025 \text{ in. (0.0064 cm)}$$

This means that the system will repeat any commanded position within ± 0.0025 in. (± 0.0064 cm) every time it is commanded. This repeatability could be improved by increasing the amplifier gain, but remember that the gain we used here is already above the calculated maximum allowable to ensure system stability.

7.3.6 Tracking Error

As long as the cylinder is moving, the cylinder will lag the position commanded by the error signal by a small distance. This *tracking error* will be at its maximum and remain constant as long as the operational amplifier output is at the saturation current, that is, as long as the valve is fully open. Below the saturation current, this value decreases until it finally reaches zero at the commanded position (at least theoretically).

Either Equation 7.41 or 7.42 can be used in this calculation:

$$T_e = \frac{I_{\text{sat}}}{G_A H} \tag{7.41}$$

or

$$T_e = \frac{v_{\text{cyl}}}{GH} \tag{7.42}$$

where GH is the open-loop gain. For our example circuit, we know that the saturation current is 200 mA. From Equation 7.41, then, the maximum tracking error is

$$T_e = \frac{I_{\text{SAT}}}{G_A H}$$

$$= \frac{200 \text{ mA}}{\left(500 \, \frac{\text{mA}}{\text{V}}\right)\left(4 \, \frac{\text{V}}{\text{in.}}\right)} = 0.1 \text{ in.}$$

To use Equation 7.42, we must first calculate the maximum cylinder velocity. In this case, the maximum flow rate was given as 5 gpm (19.2 in.^{3}/s), so

$$v_{max} = \frac{Q_{max}}{A_P} = \frac{19.2 \text{ in.}^3/\text{s}}{4 \text{ in.}^2} = 4.8 \text{ in./s}$$

This gives us

$$T_e = \frac{V_{cyl}}{GH} = \frac{4.8 \text{ in./s}}{48 \text{ s}^{-1}} = 0.1 \text{ in.}$$

(fortunately, the same value). Thus, *with the valve fully open*, the cylinder will always be 0.1 in. behind where the system thinks it should be. This error diminishes as the valve begins to close, and goes to zero as the cylinder reaches the commanded position.

7.3.7 Velocity Control Systems

Virtually everything we have said about position control systems also applies to velocity control systems, although some modifications in terminology might be necessary. The primary difference in the two systems is that a velocity control circuit requires the use of an integrating element in the operational amplifier. Therefore, a PID circuit is usually employed. The obvious reason for this is that rather than closing the valve when the commanded speed is reached, the controller must hold the valve open the appropriate amount to regulate the flow rate to the hydraulic motor.

7.4   SUMMARY

Both proportional control valves and servovalves can be used to provide rapid cycling of a fluid power actuator. In general, servovalves are the faster of the two valves, although proportional valve speeds have increased dramatically over the past few years. Although the valves themselves may be capable of cycling at several hundred hertz, the output actuator will be much slower due to the mass and inertia of the actuator, the load, and the fluid. Every system has a natural frequency. Any attempt to exceed this natural frequency is likely to cause the system to become unstable, resulting in violent, high-frequency oscillations.

Servo systems (and some proportional valve circuits) usually operate in a closed-loop mode. This means that some feedback device is used to monitor the system output (speed, position, flow, pressure, etc.) and make continuous and automatic adjustments to that output to maintain a predetermined value. The speed and accuracy with which these adjustments can be made depend on the gains in the circuit. High gains provide rapid response to very small disturbances but may lead to a very "twitchy" or even unstable system that is too sensitive for practical applications. Low gains cause the system to be slow and slug-gish and can prevent it from ever reaching the exact setpoint.

The next chapter will present some of the more commonly used feedback devices.

SUGGESTED ADDITIONAL READING

The Hydraulic Trainer: Proportional and Servo Valve Technology. 1986. Lohr am Main, Germany: Mannesmann Rexroth GmbH.

Johnson, James E. *Electrohydraulic Servo Systems.* 1977. Cleveland, Ohio: Penton/IPC.

Merritt, Herbert E. *Hydraulic Control Systems.* 1967. New York: Wiley.

Tonyan, Michael J. *Electronically Controlled Proportional Valves.* 1985. New York: Dekker.

Vickers Industrial Hydraulics Manual. 1992. Rochester Hills, Mich: Vickers, Inc.

REVIEW PROBLEMS

General

1. During a high-frequency sinusoidal oscillation, a hydraulic cylinder is moving ± 0.15 in. The command input signal is calling for ± 0.20 in. What is the amplitude response of the cylinder?

2. A flow rate of 16 gpm has been commanded from a valve. The actual output is mea-sured at 11 gpm. Find the amplitude response of the valve.

3. A hydraulic cylinder has a 4-in. bore and a 2.5-in. rod. Its stroke is 18 in., and it moves a 3000-lb resistance. It is connected to a servovalve by steel tubing that is 1 ft long (on each end of the cylinder) and has an ID of 0.66 in. The fluid's bulk modulus is 2.0×10^5 psi. The system compliance is 4.5. Find the usable frequency (Hz) of the system.

4. A hydraulic cylinder with a 3-in. bore and a 2-in. rod is used to move a 15,000-lb load. The cylinder has a 15-in. stroke. The pipes on each end of the cylinder are 14 in. long and have a 0.75-in. ID. The fluid bulk modulus is 2.5×10^5 psi. The system has a compliance of 4. Determine the usable frequency (Hz) of this system.

5. A single-ended hydraulic cylinder has a 4-in. bore, a 2-in. rod, and a 16-in. stroke. It oscillates a 4000-lb load. Each of the cylinder ports is connected to the control valve by 9 ft of 0.75-in. steel tubing. The fluid bulk modulus is 2.0×10^5 psi. Find the us-able frequency (Hz) of the system.

6. Determine the usable frequency (Hz) of a hydraulic motor used to oscillate a load with a reflected inertia of 16 lb·in·s^2. The volumetric displacement of the motor is 3.1 in.^{3}/rev. The total length of pipe connecting the motor to the valve (both sides) is 14 ft. The piping has a 1.0-in. ID. The fluid bulk modulus is 175,000 psi.

7. What would be the usable frequency of the system of Problem 6 if the valve was mounted directly on the motor?

8. Determine where the minimum natural frequency would occur for a double-ended hydraulic cylinder. Modify the cylinder circuit equation for natural frequency for use specifically with double-ended cylinders.

Express the control ratio of the following block diagrams in their simplest algebraic form:

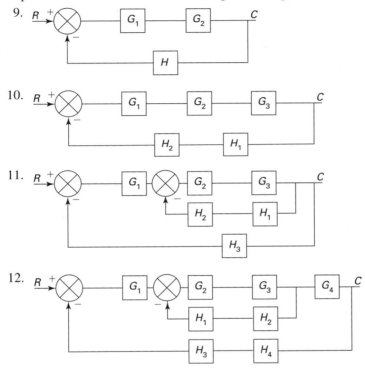

9.

10.

11.

12.

13. A joystick operates on 6 VDC. It has a throw of 24°. Find its gain.
14. A servovalve torque motor is limited to a 300 mA input current. When the valve is fully open, it passes 20 gpm. Find its gain.
15. A hydraulic motor turns at 3000 rpm when the flow rate is 50 gpm. Find its gain.
16. A hydraulic cylinder moves at 2 ft/s when the flow rate is 3 gpm. Find its gain.
17. A tachometer-generator produces 12 VDC when turned at 600 rpm. Find its gain.
18. A linear position potentiometer has a 5-in. stroke. It is powered by a 12 VDC power supply. Find its gain.
19. For the circuit block diagram shown, determine the motor speed for a command input of 4.5 V.

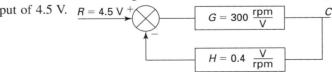

20. For the closed-loop circuit shown, determine the distance the cylinder will move from its fully retracted position if the command input signal is 4 V.

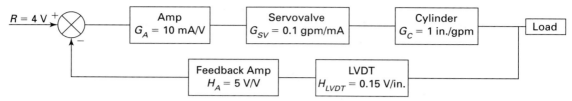

21. A servocylinder positioning system uses a command potentiometer that has 340° of rotation. The pot is powered by a 24 VDC power supply. The cylinder position is monitored by a linear potentiometer using a 12 VDC power supply. There is a 2 V/V feedback amplifier. The cylinder has a 2-in. bore, a 1-in. rod, and a 12-in. stroke. The servo amplifier produces a 100 mA output for each volt of input. The system flow rate is 10 gpm. The torque motor on the servovalve is rated at 300 mA.
    a. Draw the block diagram of the circuit.
    b. Find the control ratio (system gain) of the circuit.
    c. Determine how far the cylinder will move in response to a 35° setting of the command pot.
    d. Determine how close the cylinder will be to its final position when the valve begins to close, causing the cylinder to decelerate.

CHAPTER 8

Sensors

OBJECTIVES

When you have completed this chapter, you will be able to:

- Define sensors and transducers.
- Explain the operation of various types of sensors used as switches.
- Name and describe the operational concepts of discrete and continuous linear position sensors.
- Name and explain the operational concepts of discrete and continuous rotary position sensors.
- Discuss the operation of linear and rotary velocity sensors.
- Explain the difference between analog and digital signals and discuss the use of analog-to-digital and digital-to-analog converters.

8.1 INTRODUCTION

In the previous chapters we discussed various types of control systems in some detail. In almost all cases the control of the operation depended on some type of feedback. In those discussions little was said about the source of that feedback information. This chapter discusses the sensors and transducers that provide the signals used by PLCs and other electronic devices to control system sequencing and motion.

8.2 TRANSDUCERS VERSUS SENSORS

Although the terms *sensor* and *transducer* are often used synonymously, they are completely different. A sensor is usually part of a transducer. *Sensors* are devices that *acquire information*. The information they acquire may be related to pressure, temperature, flow rate, position, speed, or any number of other parameters.

A *transducer* is a device that converts a signal from one form to another. In our situation this usually means that the transducer receives a signal of some sort from a sensor and converts it into an electrical output that can be used by the controller. In many cases it is very difficult to distinguish between a sensor and a transducer. For example, a limit switch is a mechanical sensor used to determine that a certain position has been reached. It is also a transducer that converts mechanical motion of the sensor into an electrical signal to be used by the controller. In many similar cases, the distinction becomes blurred.

8.3   SIGNAL CONDITIONING

Signal conditioning units are devices that receive the signal from a transducer, manipulate it as necessary, and retransmit it to a controller, recorder, or display unit. The manipulation may include amplifying, filtering, converting from analog to digital, modulating, or other functions. The most commonly found devices are analog-to-digital converters.

The output of many sensors is an *analog* signal. Pressure, temperature, flow, velocity, and so forth are all analog parameters. An analog signal is one that is continuous and exactly represents the measured parameter at every instant. A *digital* signal, in contrast, is *not* continuous; rather, it progresses stepwise and does not *necessarily* represent the measured parameter at *any* time. Figure 8.1 shows the analog and digital representations of a pressure curve. Whereas the analog curve provides a smooth and continuous representation, the digital signal is in discrete steps that generally only approximate the actual pressure. Because most microprocessor-based controllers can accept only digital data, the analog output must be converted to a digital signal before it can be used. This is the function of the analog-to-digital converter (ADC).

As you can see from Figure 8.1, the digital signal is only an approximation of the analog value. How closely it can approximate the actual value is termed *resolution* or *dis-*

Figure 8.1   Analog and digital representations of a pressure curve.

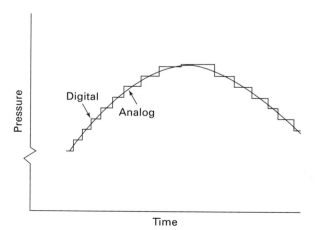

crimination. The resolution of a device defines the minimum change in the measured parameter required to produce a change in the output signal.

Microprocessor-based devices normally operate on binary code, using various combinations of 0 and 1 (representing electrical low and electrical high, respectively) to represent numbers. In the processor these binary units are fed into specific locations in the memory. These locations are organized in rows on the memory chips. The individual locations are called *bits,* and each row may contain 8, 16, or 32 bits, depending on the microprocessor used in the central processing unit (CPU). In most processors, 8 bits are grouped together to form the numbers that represent the measured parameters.

The binary, or base-2, system is conceptually the same as the decimal or base-10 system we commonly use. For example, if we write the number 6732, it tells us that there are six 1000s, seven 100s, three 10s, and two 1s that are added together to give the actual number. The only difference in the base-2 system is the value of each place. In base-2 the values for eight places are

$$128s \quad 64s \quad 32s \quad 16s \quad 8s \quad 4s \quad 2s \quad 1s$$

As you can see, each value is double the value of the place to its right rather than 10 times the value as in the base-10 system. As in the base-10 system, we add up the values shown in each place. For example, the binary number 00000111 tells us that there are no 128s, 64s, 32s, 16s, or 8s, and that there is one 4, one 2, and one 1. Adding these together (4 + 2 + 1) shows that the value of 00000111 in base-2 is 7 in base-10.

The significance of this in microprocessors is that in an 8-bit system the largest number possible is 255, and it is represented by all 1s.

$$1 \quad 1 \quad 1 \quad 1 \quad 1 \quad 1 \quad 1 \quad 1$$
$$128 + 64 + 32 + 16 + 8 + 4 + 2 + 1 = 255$$

Also, including the case of all zeros, any input must be represented by an output ranging from 0 to 255, giving a total of 256 steps.

Now, let's see how this representation affects the resolution of a system. Say we have a system to measure pressures that range from zero to 100 psi (0 to 689.5 kPa). Using the binary system, zero pressure is represented by 00000000, and 100 psi (689.5 kPa) is represented by the highest possible value (11111111) to give the best resolution. Dividing the pressure range by the possible numbers in the 8-bit system gives us a resolution of (100/256 =) 0.390625 psi (2.69 kPa). This means that a pressure increase of anything less than 0.39 psi (2.69 kPa) will not be detected by the system. Increasing the number of bits from 8 to 12 increases the number of steps to 4096, giving a resolution of about 0.024 psi (0.165 kPa).

8.4   SENSOR CHARACTERISTICS

In addition to resolution, there are several other important characteristics that are used to define sensor performance. The most important of these are accuracy and precision.

8.4.1 Accuracy

Accuracy is the ability of an instrument to give results that are the true value of the measured parameter. The accuracy of any device is established by comparing its output with a known, true standard using a standard calibration process. Any discrepancy between the instrument reading and the known value of the standard is termed the *instrument error*. This may be listed as either the *absolute error* or the *relative error*. The absolute error is the difference between the measured value and the known standard value.

It is more common to express accuracy based on the relative error, which is defined as the absolute error divided by the true value. The relative error for many devices is expressed as the percent of full scale, that is, the percent of the largest value that can be measured with the device. For example, a flow meter may have a quoted accuracy of ±5% of full scale. If the scale goes to 10 gpm (37.85 lpm), then the accuracy at 10 gpm (37.85 lpm) is ±0.5 gpm (1.89 lpm). Although this seems fairly good, a look at the other end of the scale does not look good at all, because that ±0.5 gpm (1.89 lpm) is the error *throughout the scale*. That means that if the lowest scale reading is 1 gpm (3.785 gpm), the accuracy at that point is still ±0.5 gpm (1.89 lpm), giving ±50% error!

In some cases, the relative error is expressed as a percentage of the *actual* reading. This might be expressed as ±5% throughout the scale or ±5% of the reading. In this situation the observed value is *always* within ±5% of the actual value.

False impressions of sensor accuracy can result from incorrect reading of the scale or overzealous precision in the conversion of the results. When reading scales, the smallest result that can be used is half of the smallest scale division. For example, a pressure gauge that is graduated in 20 psi increments can be read to only 10 psi (68.75 kPa). If the pointer is between 240 and 260, a reading of 250 is acceptable, but readings of 247, 255, and the like are not.

When converting values between the U.S. customary and SI systems, you must be careful not to imply a greater accuracy than was actually experienced. It is misleading to write the conversion of 25.0 in. as 63.5 cm, because the use of the single decimal place implies an accuracy of ±0.1 in. Thus, the actual conversion can properly be expressed as being somewhere between 63.246 and 63.754 cm (or 63.5 ± 0.254 cm).

8.4.2 Precision

Precision defines the instrument's ability to provide the same measurement of the same quantity under the same conditions repeatedly. Notice that precision is unrelated to accuracy. An accurate instrument is, by implication, precise, but a precise instrument is not necessarily accurate, as illustrated in Figure 8.2.

Two characteristics related to precision are repeatability and reproducibility. *Repeatability* defines the ability to produce the same results when measuring the same quantity under the same conditions over a *short time period*. *Reproducibility,* in contrast, is determined over a *long time period,* with the reading perhaps done by different people, or even using different instruments or done in different facilities.

When considering the sensors and transducers to use for a particular application, the importance of these various characteristics must be given considerable thought. In some cases, they may be extremely important. For example, the ability to consistently measure position accurately and precisely is absolutely necessary in a robot used for

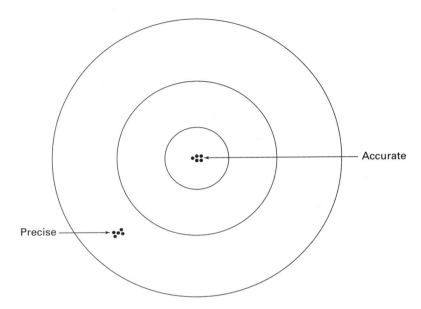

Figure 8.2 The tight pattern indicates precision. The pattern in the bullseye indicates both accuracy and precision.

assembly of small components, whereas the ability to measure the pressure in a hydraulic system to within ±100 psi (689.5 kPa) may be completely satisfactory. Buying unnecessary accuracy and precision is a waste of money.

8.5   DISCRETE SENSORS AND TRANSDUCERS

The most common sensors and transducers used for motion and sequence control are simply on-off switches that are actuated by a variety of system parameters. Such switches respond to the system parameter to complete or break the control circuit or provide an input to the PLC.

8.5.1 Limit Switches

Limit switches are used to open or close the control circuit when actuated by the presence (or absence) of some portion of the system or the load. There are several different types of limit switches: electromechanical, inductive, capacitive, light sensitive, optical. The most common type is electromechanical.

Electromechanical Limit Switches   An *electromechanical* limit switch has either a lever or a pushbutton that is mechanically actuated. Limit switches are used to indicate that the load has reached a specified position. They may be normally open or normally closed, depending on the function they are to perform. The switches are usually manufactured with three electrical connectors, one of which is the common. Whether the switch is normally open or closed is determined by which of the other connectors is used. A normally open switch is closed when actuated, thus completing the electrical circuit. A normally closed switch opens when actuated and breaks the electrical circuit. Regardless of type, these switches revert to their normal condition as soon as the actuation contact is removed.

As we discussed in Chapter 3, switches of this type are not terribly sophisticated electrically. There are, however, certain electrical aspects that must be considered whenever such electromechanical switches are applied. The most significant of these is the power-handling capacity of the contacts. Although these switches can be used in both AC and DC circuits, if the total power (watts) being switched exceeds the contact capability, the contacts are likely to be burned (which increases the resistance through the contacts and aggravates the problem) or even welded together when they are closed. The manufacturers' data provide information on the power-handling capability.

There are also mechanical considerations in limit switch applications that involve the switching characteristics of the particular switch design. These include the following (Balluff 1987):

Switching (actuating) force: The force that must be applied to fully carry out the switching operation.

Pretravel: The distance through which the actuator moves from the actuator free position to the actuator operating position.

Tolerable overtravel for constant use: The maximum allowable distance the actuator travels after reaching the switching point in constant use. Further travel could cause structural damage to the switch.

Movement differential: The distance from the operating position to the releasing position.

Actuating point: The point at which the contacts are closed (or opened). This may be adjustable in certain switches.

Releasing position: The position of the actuator at which the contacts snap from the actuated contact position to the normal contact position.

Repeatability: The ability to consistently maintain the actuating point after successive operations with constant approach axis, constant approach velocity, and constant ambient conditions, expressed in terms of plus or minus distance. Figure 8.3 illustrates the action of limit switches referred to the travel of the actuator.

Inductive Proximity Switches   Noncontact discrete position sensing can be accomplished through the use of inductive and capacitive proximity switches or by photoelectric sensors. These devices are much more sophisticated than the electromechanical switches just discussed.

Inductive proximity switches are activated by the presence of a metal within the sensing zone. These switches consist of four functional sections—the sensing head, the oscillator, the detector, and the output amplifier. The relationship of these sections is shown in Figure 8.4 (Balluff 1988).

In the operation of the switch, an inductive alternating radio frequency (RF) field is developed at the front of the sensing surface by the resonant circuit coil of the oscillator.

Figure 8.3 Electromechanical limit switch travel characteristics.

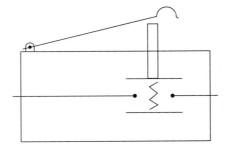

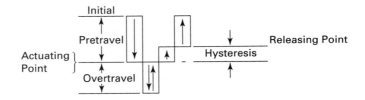

This signal projects forward from the active face of the switch. Energy is absorbed when a metal piece is brought into the RF field near the sensing surface. This causes the oscillation to stop.

The demodulator converts the oscillator signal to a DC voltage, which controls the trigger. When the oscillator signal is interrupted, the voltage feeding the trigger is also

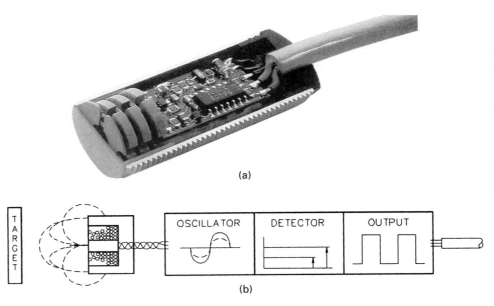

Figure 8.4 Inductive proximity switch. (a) Cutaway. (b) Block diagram of an inductive proximity switch circuit. (Courtesy of Turck Inc.)

interrupted, causing the trigger to change its condition. This results in an output from the amplifier to the control circuit.

For a DC switch, the trigger switches an output transistor (either NPN or PNP) to drive an electronic circuit or relay. In AC switches, a thyristor is triggered when the oscillation ceases. The thyristor is connected in series with a bridge. This bridge is connected to the AC load in the same way as a mechanical limit switch. The output of the sensor provides the switching signal to the control circuit. Figure 8.5 shows some typical proximity switches.

Capacitive Proximity Switches   An inductive proximity switch requires the presence of a metal to activate the trigger. In applications where there is no metal for a target, a *capacitive* proximity switch can be used. The capacitive device contains the same four functional units of the inductive sensor. In this design, however, the oscillator does not operate continuously. Rather, when an object approaches the active surface of the sensor, the capacitance increases between that object and the active zone of the sensor. When a preset value is exceeded, the oscillator starts. This produces a signal that activates the trigger, causing a change in the switched output condition. This output signal is then used by the control circuit.

Proximity switches (both inductive and capacitive) are available in a wide variety of shapes, sizes, and mounting configurations to satisfy virtually any application requirement. Because they are electronically active devices, a supply voltage separate from the control circuit is required. Each particular switch design has its specific input requirements. Proximity switches have a limited switching capacity; therefore they are seldom used to switch power directly. Rather, they are normally used to switch electronic control circuitry (PLCs, for example) or small relays that have low current requirements. Important terminology associated with these devices includes the following:

Active face: The surface where the electromagnetic or capacitive field emerges.

Operating distance: The distance at which the approaching target causes the switch output to change.

Figure 8.5 Typical proximity switches. (Courtesy of Balluff, Inc.)

Nominal operating distance, Sn: The distance at which the switch will activate when the temperature, voltage, and target are at specified test values. All switches of a particular design will switch within 10% of *Sn*. The maximum value of *Sn* is usually around 20 mm.

Effective operating distance, Sr: The actual switching distance for an individual proximity switch. It is determined under specified conditions and will be between 90 and 110% of *Sn*.

Useful operating range, S: The switching distance when the voltage and temperature are at any value within the specified range of the switch. It will be between 81 and 121% of *Sn*.

Operational distance (also termed *switching point*), *Sa:* The distance at which the operation of the switch, under stated temperature and voltage conditions, will be within 0 and 81% of *Sn* using a steel target. This value (81%) will vary when the material is not steel. Some examples follow:

Aluminum: 0.4 *Sn*

Copper: 0.4 *Sn*

Chrome–nickel: 0.9 *Sn*

Brass: 0.5 *Sn*

For capacitive switches, the switching distances are reduced to the following values when nonmetallics are sensed:

PVC: 0.6 *Sn*

Oil: 0.1 *Sn*

Wood: 0.2 to 0.7 *Sn* (depending on moisture content)

Glass: 0.5 *Sn*

Water: 1.0 *Sn*

Repeatability: The repeated effective switching distance of two consecutive switchings within 8 hours under controlled ambient and voltage conditions.

Hysteresis: The distance difference between the switching points when the target is approaching and when it is leaving the active face of the switch. The value is expressed as a percentage of the switching distance, *Sa*. Without hysteresis, the switch would "hunt" should there be any vibration of the target. Figure 8.6 illustrates hysteresis.

Most proximity switches have protective devices built into them to prevent damage to the switch from incorrect power supply polarity, back-emf from inductive loads, and short-circuit overloads.

Magnetic Reed Switches   *Magnetic reed switches* are limit switches that can be mounted directly on some (usually pneumatic) cylinders as shown in Figure 8.7. This type of switch requires a cylinder designed specifically for the application. First, a magnetic ring must be mounted on the piston. Second, the cylinder barrel must be constructed of a magnetically permeable material (usually stainless steel). As the magnet on the piston moves close to the reed switch the switch armature is attracted to the magnet, causing the switch to close and complete the electrical circuit. When the piston later moves away from the

Figure 8.6 Proximity switch hysteresis defines the difference between the switch points when the target is approaching and when it is receding. (Courtesy of Balluff, Inc.)

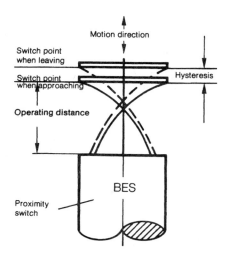

switch, the switch opens and breaks the circuit. The switch mounts are designed so they can be moved along the cylinder barrel to allow for adjustment of the end positions.

This type switch, because of its mode of operation, is not a high-precision device. Normal hysteresis of a typical reed switch will cause a repeatable error of up to 0.015 in. (0.038 cm) at both the opening and closing positions. This switch is illustrated in Figure 8.8.

Reed switches are available for use with both AC and DC circuits, although AC currents above 500 mA usually require the addition of a solid-state triac to perform the actual switching. One manufacturer (Bimba) cautions against the use of reed switches in AC circuits with devices such as light bulbs that have high inrush currents that might damage the switch. They also suggest that the switch be protected from high-voltage spikes by using a resistor-capacitor circuit in parallel with the switch. Reed switches can switch low-power circuits directly or can provide signals to PLCs.

Hall Effect Switches   *Hall effect switches* perform a function similar to that of the magnetic reed switches discussed earlier. The switches are mounted on the cylinder and require a magnetic ring on the piston. However, there is no moving mechanism to be attracted by the magnet. Instead, a solid-state element is used that is activated by the magnetic field.

The operation of a Hall effect switch is illustrated in Figure 8.9. A constant current is passed through the Hall sensor. Unless the piston magnet is directly aligned with the sensor, the current distribution is uniform, as shown in Figure 8.9a. In this case, there is no potential difference across the output. Thus, the output voltage is zero.

If the magnet is directly aligned with the sensor, the current distribution is distorted, as shown in Figure 8.9b. This distortion, known as the *Hall effect,* causes a small potential difference across the output. This small voltage is amplified by the associated electronics and is used as a control signal.

This device is well suited to interfacing with programmable controllers but cannot be used to switch circuits directly. Hall effect sensors are available as both current-sinking (NPN)

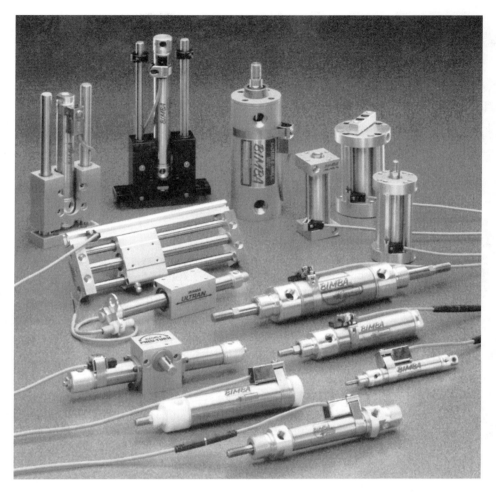

Figure 8.7 Magnetic reed and Hall effect switches are mounted externally and actuated by a magnetic ring installed on the piston. (Courtesy of Bimba Manufacturing Co.)

and current-sourcing (PNP) devices. This is an important feature, because some designs of programmable controllers require sinking inputs, and others require sourcing inputs.

Photoelectric Sensors *Photoelectric sensors* are electrical devices that respond to a change in the intensity of light falling upon them. They can be used to detect either the presence

Figure 8.8 A reed switch is a mechanical switching element that is actuated by a magnetic ring installed on the cylinder piston. (Courtesy of Bimba Manufacturing Co.)

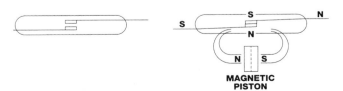

MAGNETIC
PISTON

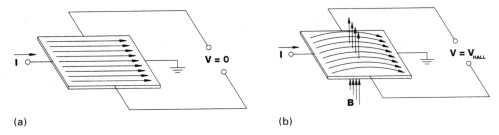

(a)   (b)

Figure 8.9 A Hall effect switch has a uniform current distribution (a) until the piston magnet causes a distortion as shown in (b). (Courtesy Bimba Manufacturing Co.)

or the absence of an object (the *target*). As a result, their applications include among others: counting, sizing, level measurement, linear and rotary speed, and timing. Another application important in fluid power circuits is that of a limit switch—determining when a specific position has been reached.

Early photoelectric sensors used a small incandescent bulb whose emitted light was focused on a photoresistive device by a collimating lens. Beginning in the early 1960s, the incandescent bulbs began to be replaced by *light-emitting diodes* (LEDs).

LEDs are solid-state semiconductors. They are similar to electrical diodes, except that they emit a small amount of light when current flows through them in the forward direction. LEDs that can emit green, yellow, blue or red light in the visible spectrum are currently available. Infrared LEDs are also available. The small amount of light emitted by the LED (much smaller than from incandescent bulbs) presents a problem from the detector end of the system. There are significant advantages, however. The first is that LEDs, being solid-state devices, will last for the entire useful life of the sensor. The second is that, since there is no filament as in an incandescent bulb, they are not as easily damaged by rough treatment.

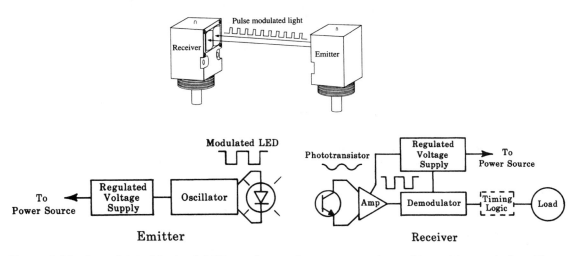

Figure 8.10 A modulated (pulsed) LED can be used to overcome the problem of low emission. (Courtesy of Banner Engineering Corp.)

Figure 8.11 The opposed sensing mode requires both an emitter and a receiver. (Courtesy of Banner Engineering Co.)

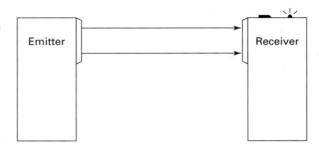

A third advantage (which actually helped overcome the disadvantage of the low emission) is the ability to switch an LED on and off very rapidly. This led to the development of the *modulated* or *pulsed* LED. The LED can be modulated at a frequency of several kilohertz. The receiver (usually a phototransistor) can be tuned to respond only to light it receives at the modulated frequency and ignore all other light. This concept is illustrated in Figure 8.10. This means that, up to a limit, a modulated LED can be used under ambient light conditions. However, very bright ambient light or direct sunlight can *overpower* the receiver. This condition is termed *ambient light saturation*.

There are three distinct modes of operation for photoelectric sensors: opposed, retroreflective, and proximity. The *opposed mode* (often called the *beam-break* mode) was the first application and is still the most commonly used arrangement. This mode, illustrated in Figure 8.11, uses an emitter and a receiver set opposite each other so that the emission from the emitter impinges directly upon the receiver. Whenever the beam is interrupted, the receiver electronics produce a signal indicating that interruption.

Alignment of opposed sensing photoelectric sensors is critical: the energy from the emitter must be centered on the receiver. In early instruments, this alignment process was difficult and time consuming. Modern high-powered, modulated devices are much easier to align than the early devices. This process is a simple matter if the emitter is a visible laser.

The *sensing range* defines the maximum distance between the emitter and the receiver. Depending on the specific unit design, this range may be a few inches or as much as 300 ft (91.4 m). Throughout this range, the *effective beam* of the emitter (as illustrated in Figure 8.12) must be *completely* interrupted for reliable sensing. Often, the effective

Figure 8.12 The effective beam is only that part of the emitted beam that lies within the field of view of the receiver. (Courtesy of Banner Engineering Corp.)

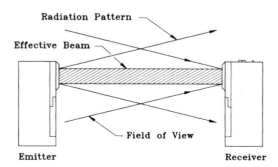

Figure 8.13 The retroreflective sensing mode requires a retro target that reflects the emitted beam back to the integral receiver. (Courtesy of Banner Engineering Corp.)

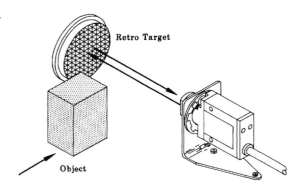

beam size is too large to reliably detect small objects. In such cases, an *aperture* can be fitted to the lens of the emitter, the receiver, or both. This is simply a lens cover with a small slit or hole that effectively reduces the size of the beam transmitted or received. The amount of light energy passing through the apertured lens is reduced by the same amount as the effective lens area is reduced.

In the *retroreflective* mode (also called the *reflex* mode or the *retro* mode), a target is used to reflect the light beam back to a single unit that contains both the emitter and the receiver. This concept is illustrated in Figure 8.13. As with the opposed mode, an object interrupting the beam triggers an appropriate output.

The targets used for this mode are called *retroreflectors* or *retrotargets*. They are usually made up of tiny corner-cube prisms such as the one illustrated in Figure 8.14. Light entering a corner-cube prism is reflected by the perpendicular surfaces and exits the prism parallel to the incident path. These are the same devices that are used in reflective markers on roadways, signs, reflective tape, and the like.

Whereas the opposed mode sensors have no minimum range, retroreflective devices cannot detect objects closer than some unit-specific minimum distance, because at close range the interrupting object may reflect enough of the beam to fool the sensor. Depending on the type unit, this minimum range may be as little as 1 in. (2.54 cm) or as much as 10 ft (3.048 m). The maximum range for specific devices varies from a few inches to as much as 100 ft (30.48 m).

As you might suspect, the sensing of highly reflective objects can be a little tricky with retroreflective sensors. In such cases, it is usually necessary to arrange the system

Figure 8.14 A corner-cube prism reflects a light beam parallel to the incoming (incident) light beam. (Courtesy of Banner Engineering Corp.)

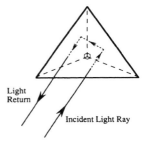

Figure 8.15 The diffuse mode uses the light reflected from the target to trigger a response from the sensor. (Courtesy of Banner Engineering Corp.)

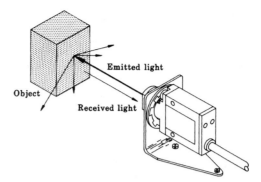

so that the light strikes the object at an angle that will not reflect it directly back to the receiver.

The *proximity sensing mode* involves using the reflection from the object itself to trigger an appropriate output. In this mode, there is no target, as in the retroreflective mode. Because of the diversity of object sizes and surface reflectivities as well as environmental conditions under which these devices can be used, there are four different optical arrangements used for proximity sensors.

The first of these is the *diffuse mode*. The diffuse mode sensor is the most common of the proximity sensors. In these devices, a collimated light beam is directed onto the surface of the object, as shown in Figure 8.15. This incident beam is scattered (or diffused) by the surface, and only a small portion is reflected back to the sensor. The detection ability of diffuse sensors is highly dependent on the reflectivity of the surface and the angle at which the incident light strikes the surface. Bright, highly reflective objects are more accurately detected and at greater ranges than dull, dark surfaces. Because the sensor detects only light that is returned parallel to the emitted light, curved or angled surfaces are difficult to detect. The range of this type of sensor is up to 6 ft (1.83 m).

A second type of proximity sensor uses the *divergent mode,* as shown in Figure 8.16. This type device is specifically intended for shiny objects and short ranges. The emitter has no collimating lens, so the sensing range is shorter. The advantage is that the unit is much less dependent on the angle of incidence of the emitted beam. The range is very dependent on the size and shape of the illuminated object. Divergent mode sensors work

Figure 8.16 Divergent proximity sensors are intended for shiny objects and short ranges. (Courtesy of Banner Engineering Corp.)

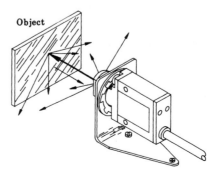

Figure 8.17 The convergent sensing mode is used to detect very small objects passing through the converging beam near the focal point. (Courtesy of Banner Engineering Corp.)

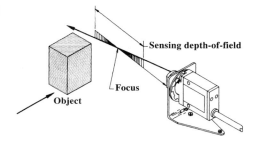

very well at ranges down to about 1 in. (2.54 cm) and can be used successfully to detect small objects, including yarn and wire.

Another type of proximity sensor uses a *convergent beam mode*. This mode is especially useful for detecting very small objects. It uses a lens system that focuses the emitted beam at an exact point, as shown in Figure 8.17. This is defined as the *focal point* or *focus*. The focal point also defines the range of the sensor. Only objects that fall within a specified distance before and after the focal point will be detected. This zone of sensitivity is the instrument's *depth of field*. Because this depth of field can be made very small, objects with very small profiles or low reflectivity can be reliably detected. Typically, focal points range from 1.5 to 6 in. (3.8 to 15.24 cm) from the emitter, with a depth of field around ±50% of the focal point.

A unique type of proximity sensor uses *background suppression*—also called *fixed-field* sensing. These devices ignore objects outside their sensing range, even though those objects may be highly reflective. To accomplish this task, two detectors are set at different angles from the emitter as shown in Figure 8.18. The electronics compare the amount of light seen by each element. Target detection occurs when the amount of light received by receiver 2 is equal to or greater than that received by receiver 1.

Figure 8.18 Fixed-field sensing is used to suppress the effects of a bright background. Object A will be sensed if the amount of light received at R_2 is equal to or greater than that at R_1. (Courtesy of Banner Engineering Corp.)

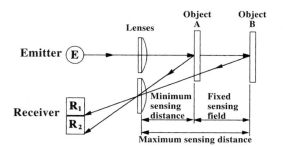

Regardless of the type of photoelectric sensor used, the device can be set to produce an output signal (switched ON or OFF) in response to either the receipt of a light signal or the absence of the light signal. These conditions are termed *light operate* (in which the output is energized when light is received) and *dark operate* (in which the output is energized when no light is received). These concepts are illustrated in Figure 8.19.

Example 8.1: Individual-size cereal boxes move along a conveyor belt past a convergent beam photodetector at the rate of 1000 per minute. The boxes are 2.5 in. (6.35 cm) wide

Figure 8.19 The opposed mode is the most common arrangement of photoelectric sensors. They can be set to operate when the receiver is receiving light (light operate) or when the beam is broken (dark operate). Courtesy of Banner Engineering Corp.)

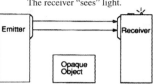

LIGHT OPERATE

The output is energized when the beam is unblocked.
The receiver "sees" light.

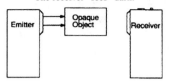

DARK OPERATE

The output is energized when an object blocks the light from reaching the receiver.
The receiver "sees" dark.

and are evenly spaced at 0.5-in. (1.27-cm) intervals. The photodetector provides information to a process controller so that 12 boxes at a time are separated and packaged. What is the minimum sensor response time required to accurately count the boxes?

Solution: First, we need to determine the speed of travel of the boxes. The width of the boxes plus the separation means that there is a total travel of 3.0 in. (7.62 cm) from the leading edge of one box to the leading edge of the next. Thus, the speed of the conveyor is

$$\left(3.0 \ \frac{\text{in.}}{\text{box}}\right)\left(1000 \ \frac{\text{boxes}}{\text{min}}\right)\left(\frac{\text{min}}{60 \ \text{s}}\right) = 50 \ \frac{\text{in.}}{\text{s}} \left(127 \ \frac{\text{cm}}{\text{s}}\right)$$

The time that each box is in the beam and seen by the detector is

$$t = \frac{2.5 \ \text{in.}}{50 \ \frac{\text{in.}}{\text{s}}} = 0.05 \ \text{s} = 50 \ \text{ms}$$

Before we start looking for a sensor with a 50-ms response time based on the box *width,* we need to make the same calculations based on the box *separation.* Here, we find that

$$t = \frac{0.5 \ \text{in.}}{50 \ \frac{\text{in.}}{\text{s}}} = 0.01 \ \text{s} = 10 \ \text{ms}$$

This result tells us that the *maximum* acceptable response time is 10 ms. We would leave a little cushion by choosing a detector with a faster response time than the calculations indicate.

A common application for retroreflective sensors is in determining velocity by detecting the pulses from reflective strips mounted on a moving object (target). This is especially effective for rotating-shaft applications in which a small piece of reflective tape is affixed to the shaft. The following example illustrates this concept.

Example 8.2: A retroreflective detector with a 4-ms response time is used to monitor the rotary speed of a hydraulic motor with a 3-in. (7.62-cm) shaft. A piece of reflective tape

1 in. (2.54 cm) long is used as the target. Determine the maximum speed that can be accurately measured by the detector.

Solution: The concepts involved here are the same as in the previous example except that we must convert between linear and rotational speeds. Based on the 4-ms response time of the sensor, the maximum *linear* velocity of the 1-in. piece of tape that can be accurately measured is

$$\text{Linear velocity} = \frac{\text{target length}}{\text{response time}} = \frac{1 \text{ in.}}{4 \text{ ms}} = \frac{1 \text{ in.}}{0.004 \text{ s}} = 250 \frac{\text{in.}}{\text{s}} \left(635 \frac{\text{cm}}{\text{s}}\right)$$

The circumference of the shaft is

$$c = \pi d = 3\pi = 9.42 \text{ in. (23.93 cm)}$$

This is the linear travel on the shaft circumference per revolution, so we can express this result as 9.42 in./rev. The maximum rpm, then, is

$$\text{rpm} = \frac{\text{tape linear velocity}}{\text{shaft linear velocity}} = \frac{\left(250 \frac{\text{in.}}{\text{s}}\right)\left(\frac{60 \text{ s}}{\text{min}}\right)}{9.42} = 1592 \text{ rpm}$$

If the shaft speed exceeds 1592 rpm, the detector cannot switch fast enough to measure the speed accurately. In simple applications such as this one, it is the usual practice to make the length of the target half the circumference of the shaft. Doing so in this case would increase the maximum detectable speed to 7500 rpm.

Ultrasonic Sensors   *Ultrasonic sensors* are similar to photoelectric sensors, but they emit and detect sound waves instead of light. The sound frequencies used are above 20 kHz, which is the upper range of human hearing. These devices can be used in both the opposed and the proximity modes. The receivers are tuned to filter out sound at frequencies other than that of the transmitter. The receiver (and sometimes the emitter) is often fitted with a *waveguide* to shape the ultrasonic beam and eliminate echoes from objects to the side of the beam path. The range of ultrasonic sensors depends on environment and target size but generally runs from a few inches to about 20 ft (6.1 m).

8.5.2 Pressure Switches

Pressure switches are used to detect prescribed pressure situations and switch the control circuit in response to those situations. They can be adjusted to respond to high pressure, low pressure, or even vacuum. The response to the sensed pressure is the changing of the state of the contacts to switch or initiate an output signal that can be used to switch a control device.

There are numerous pressure switch designs, including Bourdon tube, bellows, piston, and diaphragm mechanisms. In all cases the movement of the mechanism operates the electrical contacts in the switch. The switch can be normally open or normally closed. In fact, most switches are delivered with three electrical leads. One is attached to the common contact. The selection of one of the other two leads determines whether the switch is normally open or closed.

Pressure switches are often described by the terms *make on rise* and *break on rise*. These terms describe the operation of the contacts in response to increases in pressure.

There are both mechanical and electrical considerations in the selection and use of pressure switches. Mechanical considerations are the operating pressure, adjustment range, proof pressure, burst pressure, and pressure differential.

The switching characteristics of the pressure switch illustrated in Figure 8.20 are related to these mechanical considerations. The *proof pressure* is the maximum pressure to which the switch can be subjected without damage to the switch mechanism and the body. The *burst pressure* (usually two to four times the proof pressure) is the pressure at which the switch body will experience structural failure. The *pressure differential* (the difference between the switch point and the switch-back or reset point) is adjustable on some switches and can be used to adjust the on-off cycle rate of the system.

Electrically, pressure switches are very similar to any other contact-type switch, so the same terminology and limitations apply. As with all switches, the power-handling capacity of the contacts must not be exceeded.

Because pressure switches sense and respond to pressure, they can be used to provide the operating signal for pressure and directional control valves as well as for PLCs. One fairly common application is to use a two-way, two-position, normally closed single solenoid valve as an unloading valve. When the pressure reaches the pressure switch setting, the solenoid on the unloading valve is energized, opening the valve and dumping system pressure. As another example, pressure switch signals to PLCs can be used to indicate that a desired (or undesired) pressure limit has been reached or to provide an emergency stop signal.

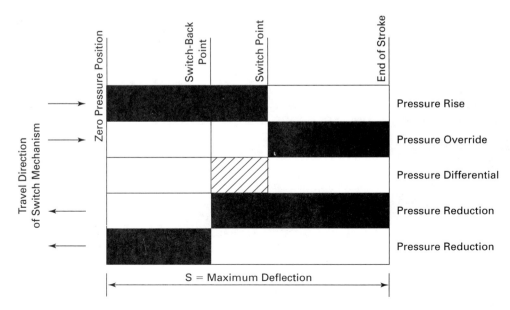

Figure 8.20 Switching characteristics of a typical pressure switch.

Pressure switches automatically reset themselves in the same manner as do mechanical limit switches. As a result, a holding circuit must be used if the control circuit is required to maintain the solenoid in its switched position.

8.5.3 Other Discrete Elements

In addition to position and pressure, switching devices are available that respond discretely to other system parameters such as flow rate, flow volume, liquid level, hydraulic motor speed, and fluid or hardware temperature. Mechanically, each of these devices is obviously very different. Electrically, however, they normally employ a simple contact pair with the same characteristics and limitations as those in any other switching device.

8.5.4 Timers and Counters

Although they are not sensors per se, timers and counters play an important role in many industrial processes. Timers may be electronic or mechanical. We will limit this discussion to electronic timers.

An electronic timer is often called a time-delay relay, because its operation is similar to that of a standard relay. Like relays, a timer usually controls sets of contacts that it opens and closes. The difference is that the timer contacts do not switch until a designated time has elapsed. Depending on the application, the time delay can follow either the energizing or deenergizing of the timer. Some electronic timers can be set in intervals as small as 0.01 s.

Counters used as control elements operate in much the same way as do timers. In most programmable logic controllers, the same elements and addresses are used for both timers and counters. The only real difference is the input signal to the element. The timer receives an input from an internal clock, whereas a counter receives its input as pulses from some external device such as a limit switch, proximity switch, or photoelectric sensor. After receiving a preset number of pulses, all contacts associated with the counter are switched. Programmable logic controllers are discussed in detail in Chapter 9.

8.6   CONTINUOUS SENSORS

Unlike discrete sensors which are used to detect individual events, continuous sensors are active throughout an operation, continuously monitoring a parameter and producing an output representing that parameter. Such devices include temperature, pressure, and flow rate sensors as well as units that sense linear and rotary position and velocity. The outputs from these devices can be displayed to indicate status or used as feedback and control signals. In the following sections, we will discuss devices for continuously measuring temperature, flow rate, pressure, linear position, and rotary position.

8.6.1 Temperature Sensors

Fluid temperature is seldom a major operating parameter in fluid power systems; however, it is often necessary to know when a specific temperature limit is reached. These limits (either high or low) may be used to indicate whether the oil is overheating or if it is too cold. The operation of these sensors may be transparent to the operator, which means that they may cause some change in the machine operation without the operator's knowledge. For example, if the fluid is too cold, the sensor may turn on a fluid heater. If it is too hot, it may open a valve to divert flow through a cooler. Either response may simultaneously turn on a light or annunciator on the control panel to let the operator know that the action has been taken. These devices are usually simple temperature switches that use bimetallic elements or other temperature-sensitive devices to open and close contacts.

In some situations, however, the actual fluid temperature is of interest. In such cases, sensors are required that will provide a constant temperature readout. A variety of devices can be used, the most common of which are thermocouples, thermistors, and resistance temperature detectors (RTD).

A *thermocouple* consists of two wires of dissimilar metals joined at the ends. When there is a temperature difference between the two junctions, an electrical potential is created. This phenomenon is called the *thermoelectric effect* (or the *Seebeck effect* after its discoverer). This potential causes a current to flow around the circuit, as shown in Figure 8.21. A measurement of the electrical potential (voltage difference) between the two junctions can then be translated into a temperature differential. If the temperature of one junction is known, the differential can be used to determine the temperature at the other junction.

The current flow in the loop depends on several factors—the two metals, the wire size, the wire length, and the absolute temperatures at each junction, in addition to the temperature differential. Interestingly, intermediate temperatures along the wires do not affect the results—only the temperatures of the junctions are of concern. The response to the temperature differential is nonlinear and depends on the absolute temperatures of the junctions. For example, if the individual junction temperatures are 1500 R (833 K) and 500 R (278 K), the differential is 1000 R (556 K) and will produce a corresponding voltage. If the temperatures are 1600 R (889 K) and 600 R (333 K), the differential is still 1000 R (556 K), but the resulting voltage will be different. Because of this nonlinearity, temperature compensation is usually included as a feature of the signal conditioning unit.

Virtually any two dissimilar metals can be used for the thermocouple wires, but experience, tradition, and standardization have narrowed the combinations to a few that have proved to work well over the full range of common applications and temperatures.

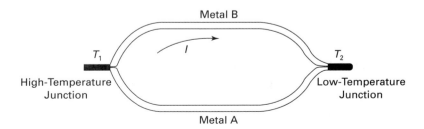

Figure 8.21a The Seebeck thermoelectric effect results when the junctions of two wires of dissimilar metals are exposed to different temperatures.

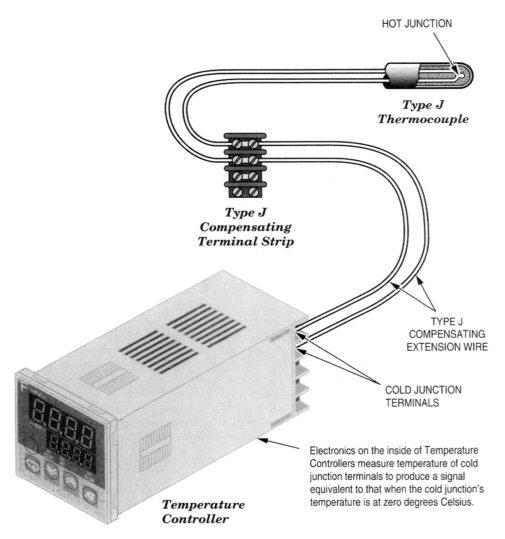

HOT JUNCTION

Type J
Thermocouple

Type J
Compensating
Terminal Strip

TYPE J
COMPENSATING
EXTENSION WIRE

COLD JUNCTION
TERMINALS

Electronics on the inside of Temperature
Controllers measure temperature of cold
junction terminals to produce a signal
equivalent to that when the cold junction's
temperature is at zero degrees Celsius.

Temperature
Controller

Figure 8.21b The Seebeck effect can be converted into digital readout using compensating electronics. (Courtesy of Womack Machine Supply)

Although the wire junction itself constitutes the thermocouple, the wire usually needs to be supported and protected. As a result, commercially available thermocouples are usually encased in a protective sheath. There are three common configurations, which are shown in Figure 8.22. The ungrounded (or insulated) junction in Figure 8.22a is electrically insulated and completely isolated from any harmful elements in the environment. This protects the thermocouple but slows its response time. The grounded junction in 8.22b still protects the junction but gives it intimate contact with the thermal environment. Thus, the response time is much faster than that of the insulated junction. Figure 8.22c shows

Figure 8.22 Thermocouple junctions may be either completely enclosed (a and b) or exposed (c). Courtesy of Womack Machine Supply)

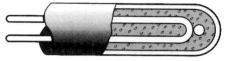

Ungrounded Junction

(a)

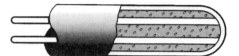

Grounded Junction

(b)

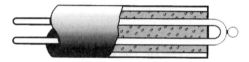

Exposed Junction

(c)

an exposed junction. Although this type has the fastest response time, it can be used only where the environment will not damage the wires.

Thermistors are low-cost semiconductor devices that have a very large change of electrical resistance in response to a temperature change, but unfortunately, this change is very nonlinear. Consequently, the electronics must include a look-up table in the memory so that an observed resistance can be converted to the correct temperature. Thermistors are normally used to measure temperatures over a limited range [usually not more than 0–100 °C (32–212 °F)] and are commonly used in temperature-alarm and switching circuits.

A *resistance temperature detector* (RTD) works on the principle that the electrical resistance of a metal wire changes with temperature. Although these devices are not as sensitive to temperature changes as thermistors, the change is much more linear. This makes them much more convenient from a signal conditioning consideration. As with thermocouples, virtually any metal can be used, but the most common are nickel, tungsten, copper, and platinum. Platinum-based RTDs are the most expensive, but they are also the most widely used because of their linearity over a wide temperature range.

RTDs lend themselves to a variety of shapes—flat film, disks, cylinders, etc.—to accommodate a variety of applications and installations. In all applications, however, the wire must be protected from stresses that would cause a change in resistance. The RTD must also be protected from physical damage and hostile atmospheres. A variety of electronic circuits utilize the RTD signal, but the most common uses the RTD as a leg of a Wheatstone bridge, as shown in Figure 8.23.

Figure 8.23 The most common types of resistance temperature detectors (RTDs) use a Wheatstone bridge.

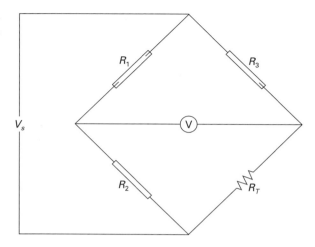

8.6.2 Flow Rate Measurement

Flow rate can be measured in a number of ways, but the most useful device for use in control systems is the *turbine flow meter*. A typical turbine flow meter is shown in Figure 8.24. It consists of the turbine, the magnetic sensor, and the bearing supports. The turbine is ferrous material (usually a magnetic grade of stainless steel) that is rotated by the fluid flow. As the blades of the turbine enter the magnetic field, shown in Figure 8.25, the

Figure 8.24 Components of a turbine flow meter. (Courtesy of Hedland Division of Racine Federated)

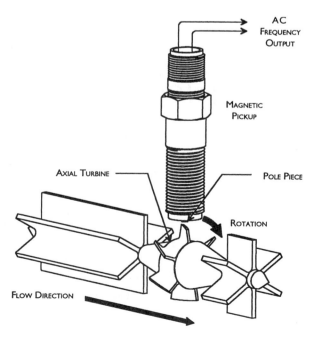

Figure 8.25 The operation of a turbine flow meter is similar to that of an AC generator. The rotating ferromagnetic turbine interrupts the magnetic field, producing an AC voltage. (Courtesy of Hedland Division of Racine Federated)

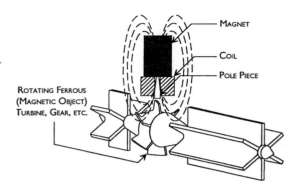

losses in the field path decrease and the magnetic field increases. The voltage output of the transducer is proportional to the amount and rate of the change in the magnetic field.

Generally, the turbine, sensor/transducer, and readout unit are purchased as a matched, calibrated set and are not interchangeable. Figure 8.26 shows a complete unit. The unit is normally calibrated for use with a fairly specific fluid or type of fluid; therefore, application in a different fluid can lead to errors. The range over which the flow meter output is linear must be observed. The nonlinearity problem is much more pronounced at low flow rates. One of the causes of this nonlinearity is the rate at which the magnetic field recovers after turbine blade passes out of it. The second is that the flow-generated forces on the turbine are not great enough to cause the turbine to rotate smoothly due to the magnetic coupling between the pole piece and the blade. At low flow rates, the turbine motion may be erratic, and it may stall completely.

Figure 8.26 The turbine flow meter, sensor, and read-out unit are usually purchased as a matched, calibrated set. (Courtesy of Hedland Division of Racine Federated)

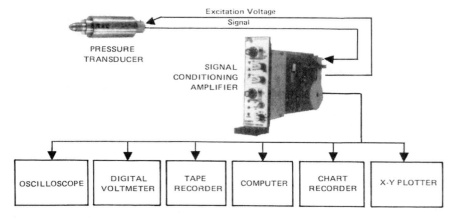

Figure 8.27   Typical pressure measuring system. (Courtesy of Omegadyne, Inc., Sunbury, Ohio)

8.6.3 Pressure Measurement

A typical pressure measurement system contains a pressure transducer, a signal conditioning amplifier, and one or more readout units, one of which may be a computer used to control the operation of a motion system. This arrangement is illustrated in Figure 8.27. Let's look briefly at the more commonly used types of pressure transducers.

The most frequently found pressure transducer is the *strain gauge* type. In a strain gauge transducer a force or displacement causes a change in the position or length of the sensing element. There are four types of strain gauge sensing elements: unbonded, bonded, thin-film, and diffused semiconductor. An *unbonded strain gauge* is shown in Figure 8.28. In this unit the stress in the strain wire varies as the spring member is deflected by pressure acting on the diaphragm.

Figure 8.28   Unbonded strain gauge. (Courtesy of Omegadyne, Inc., Sunbury, Ohio)

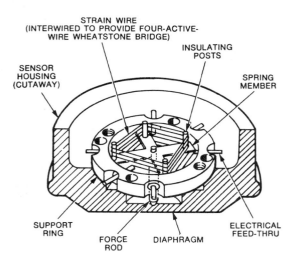

Figure 8.29 Bonded foil strain gauge. (Courtesy of Omegadyne, Inc., Sunbury, Ohio)

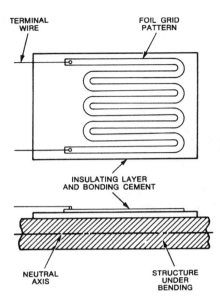

In a *bonded strain gauge* a wire or foil ribbon is bonded to a diaphragm, as shown in Figure 8.29. Pressure acting on the diaphragm causes it to flex, resulting in a stressing of the wire or film. A *thin-* (or *sputtered-*) *film strain gauge* is similar in concept to the bonded unit, but here, a thin film [less than 0.0005 in. (0.0013 cm)] of silicon dioxide is sputter-deposited on a highly polished metal diaphragm, followed by a sputter coating of the gauge material. After several intermediate steps of etching, cleaning, and such, gold wires are attached to nickel–chromium pads on the gauge material, as shown in Figure 8.30. As with the bonded strain gauge, the flexing of the diaphragm produces the stress of the gauge element. Figure 8.31 shows a complete strain gauge transducer assembly.

The final type is the diffused *semiconductor gauge*. In this device a silicon chip is used as the substrate. An impurity—usually boron—is *diffused* into it to produce a piezoresistive element. This type of device has a very high gauge factor, meaning that it can produce a relatively large signal from a relatively small strain, making it very sensitive to small changes. Its high output reduces signal distortion due to system electrical noise. It also has better overall performance, stability, linearity, and hysteresis than other strain gauge devices.

Figure 8.30 Strain gauge resistor deposited on substrate. (Courtesy of Omegadyne, Inc., Sunbury, Ohio)

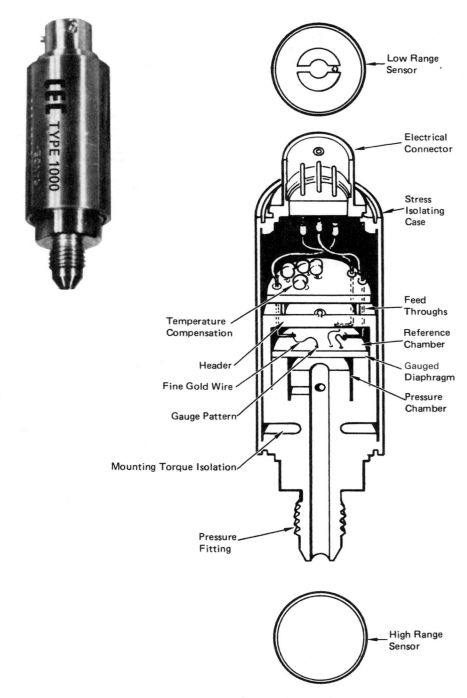

Low Range
Sensor

Electrical
Connector

Stress
Isolating
Case

Feed
Throughs

Temperature
Compensation

Reference
Chamber

Header

Gauged
Diaphragm

Fine Gold Wire

Pressure
Chamber

Gauge Pattern

Mounting Torque Isolation

Pressure
Fitting

High Range
Sensor

Figure 8.31 Cutaway view of sputtered strain gauge final assembly. (Courtesy of Omegadyne, Inc., Sunbury, Ohio)

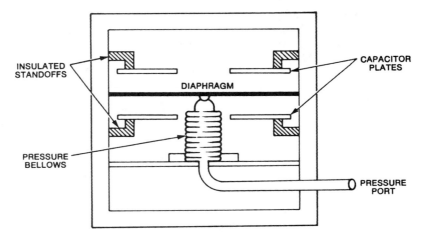

Figure 8.32 Capacitance sensor. (Courtesy of Omegadyne, Inc., Sunbury, Ohio)

A different concept in pressure measurement is that of the *capacitive sensor*. Such a device is shown in Figure 8.32. Here, pressure causes the bellows to move the diaphragm relative to the two insulated capacitor plates. The capacitance change between the two circuits is used to determine the system pressure.

A *linear variable differential transformer* (LVDT) is shown in Figure 8.33. In this pressure transducer the system pressure deflects the diaphragm (or sometimes a Bourdon tube), which moves the LVDT core. This results in a signal proportional to the pressure.

The *piezoelectric pressure transducer* operates on the physical concept that strain applied to an asymmetrical crystalline material such as quartz or tourmaline generates an electrical charge. This charge is very small; therefore, an amplifier is usually installed very near the crystal, and a low-noise cable is required between the crystal and the amplifier.

Figure 8.33 Differential transformer. (Courtesy of Omegadyne, Inc., Sunbury, Ohio)

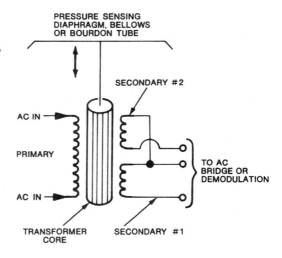

Figure 8.34 Piezoelectric pressure transducer. (Courtesy of Omegadyne, Inc., Sunbury, Ohio)

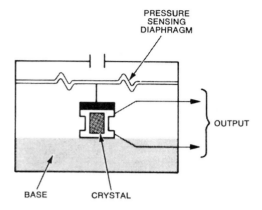

Figure 8.35 A linear potentiometer is a simple and inexpensive device for measuring linear position.

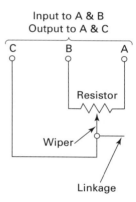

The crystals are also very sensitive to temperature changes. However, this type of transducer has a higher frequency response than any other pressure-measuring device. The general arrangement of a piezometric sensor is shown in Figure 8.34.

8.6.4 Linear Position

Linear position feedback is an extremely important factor in the control of automated equipment and in robotics. It can also play an important role in certain inspection and measurement applications. There are several devices that can be used for linear position feedback from fluid power cylinders (as well as from many other machine elements).

The simplest and least expensive device is a *linear potentiometer* or *pot,* as it is commonly called. Like any other potentiometer, this device consists of a tightly packed coil of wire wound around a supporting core. The ends of the coil are connected to a known voltage potential. A third lead is connected to a *wiper,* as shown in Figure 8.35. The wiper, which moves with the cylinder rod, travels along the coil and taps off a voltage that is proportional to its distance along the coil. The voltage output at any point is defined by

$$V = \frac{kd}{R} \times E$$

where V = voltage output
        k = constant that is characteristic of the specific potentiometer (ohms per unit length)
        d = distance from the zero end
        R = the total resistance of the coil
        E = source voltage

In most cases, this equation reduces to

$$V = \frac{d}{L} \times E$$

where L is the total length of the potentiometer

The entire device may be internal to the cylinder, or it may be externally mounted with the wiper made to travel with the cylinder rod.

Potentiometers are inexpensive and have the advantages of producing a high output without amplifiers and of being capable of operation with either AC or DC sources. On the negative side, they have coarse resolution, tend to have a limited life, develop dead spots, have low frequency response, and have a large hysteresis.

A *linear resistive transducer* (LRT) is a potentiometric device that operates much like a linear potentiometer. The LRT probe, which is built into the cylinder rod, has a resistive element on one side and a collector strip on the other, as shown in Figure 8.36. As the piston moves, an electrical circuit is created between the resistive element and the collector strip. The result is a continuous analog voltage output that is proportional to the cylinder position. The LRT has infinite resolution; therefore, the displayed resolution is a function of the analog-to-digital converter rather than the transducer. As discussed earlier, the electronic resolution is the span of the measured parameter divided by the number of steps of the converter. For example, when a 10 VDC LRT is used with a cylinder that has

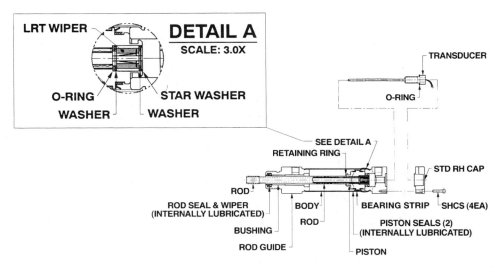

Figure 8.36 A linear resistive transducer (LRT) is similar to a linear potentiometer. (Courtesy of Bimba Manufacturing Co.)

a 10-in. (25.4-cm) stroke, the voltage is zero when the cylinder is fully retracted and 10 VDC when it is fully extended. The span, therefore, is 10 VDC. A 12-bit processor has 4096 steps. Thus, the smallest detectable increment is (10 VDC ÷ 4096 =) 2.4 mV. Since the gain of the LRT is (10 V/10 in. =) 1.0 V/in. (0.39 V/cm), this corresponds to (0.0024 V/1.0 V/in. =) 0.0024 in. (0.0061 cm). If the same LRT is used with a 12-in. (30.5-cm) stroke, the converter resolutions will still be 2.4 mV, but the LRT gain will be (10 V/12 in. =) 0.8333 V/in. (0.328 V/cm). Here, the smallest detectable increment will be (0.0024 V/0.83333 V/in. =) 0.00288 in. (0.0073 cm). Thus, the longer the stroke, the poorer the resolution.

A *magnetostrictive linear displacement transducer* (MLDT) consists of a nonmagnetic stainless steel tube (or probe) welded to an enclosed stainless steel base called the head. The probe contains a stretched magnetostrictive wire. The head contains the control electronics. A permanent magnet ring slides over the outside of the probe and serves as a movable position marker, as illustrated in Figure 8.37. This entire unit is mounted inside the rod of the cylinder, as shown in Figure 8.38, with the magnet ring attached to the inside of the cylinder rod.

The electronics in the head periodically send a short-duration current pulse along the wire. When the magnetic field generated by the current pulse intersects the magnetic field of the magnetic ring, magnetostriction causes a torsional strain on the wire. This strain travels back down the wire and is detected by the electronics. The time lapse between the emission of the pulse and the receipt of the strain signal is a precise indication of the position of the magnet ring, hence the cylinder rod. MLDTs are available in lengths up to 120 in.

Another device for determining linear position is the LVDT that we discussed earlier. Although normally considered short-stroke devices, LVDTs can provide measurements

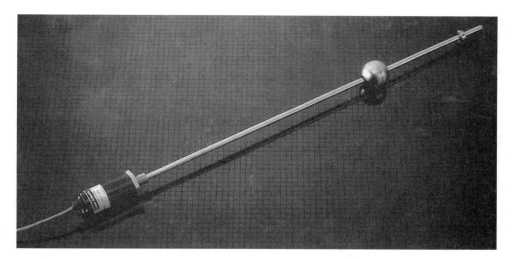

Figure 8.37 A magnetostrictive linear displacement transducer (MLDT) can be used to measure both position and velocity in hydraulic and pneumatic cylinders. (Courtesy of Lucas Control Systems, Schaevitz™ Sensors)

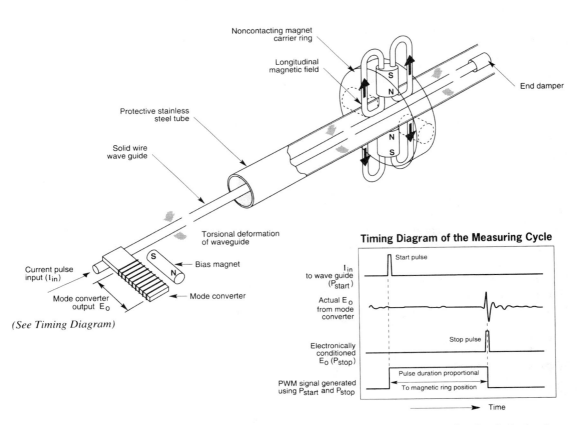

Figure 8.38 An MLDT installed in a hydraulic cylinder can provide a closed-loop feedback device for both position and velocity. (Courtesy of Lucas Control Systems, Schaevitz™ Sensors)

of strokes up to 10 ft in length. Differential transformers normally require an AC excitation; however, DC-powered LVDTs are available. The units consist of standard AC LVDT plus an integral carrier generator/signal conditioning unit. The device uses an external DC source (battery or power supply) that it converts to the necessary AC signal.

8.6.5 Rotary Position Measurement

In many machines it is necessary to have feedback for tracking the position of rotating elements. This rotary position measurement can be accomplished with a variety of devices. The most common of these are rotary potentiometers, resolvers, encoders, synchros, and rotary variable differential transformers (RVDTs), which are based on the LVDT concept.

Figure 8.39 illustrates the concept of a *rotary potentiometer*. Like a linear pot, it consists of a tightly packed coil and a wiper or brush. The output is an analog voltage that is proportional to the angle of rotation. The output is defined mathematically as

$$V = \frac{k\theta}{R} \times E$$

where
V = output voltage
k = a constant that is characteristic of the specific potentiometer (ohms per degree)
θ = angle of rotation from the zero point (degrees)
R = total resistance of the coil
E = source voltage

In most cases, this equation reduces to

$$V = \frac{\theta}{\theta_T} \times E$$

where θ_T is the total rotation of the pot in degrees

Although pots are capable of continuous rotation, they are more commonly used to measure angular displacement of less than 360°.

There are two types of encoders—incremental and absolute. An *incremental encoder* is one of the simplest rotary position measuring devices. It consists of a pair of photosen-

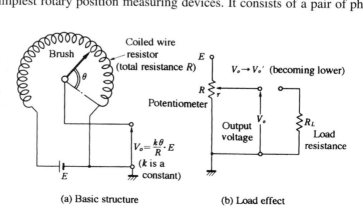

Figure 8.39 Rotary potentiometer. (Source: Intelligent Sensor Technology, edited by Ryoji Ohba. © 1992 by John Wiley & Sons Limited. Reproduced by permission of John Wiley & Sons Limited)

(a) Basic structure

(b) Load effect

Figure 8.40 An incremental encoder uses a pair of photo-sensors and a disk with two rows of slits. (Source: Intelligent Sensors, edited by Ryoji Ohba. © 1992 by John Wiley & Sons Limited. Reproduced by permission of John Wiley & Sons Limited)

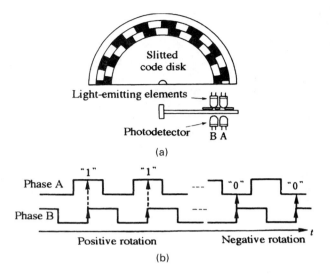

sors and a code disk with slits cut in it, as shown in Figure 8.40. As the disk rotates, the beam from the light emitters is interrupted. These interruptions are detected by the photodetectors. Each interruption causes a pulse to be generated that is used by the computer to determine an *incremental* angular displacement, that is, how far the device has rotated from its initial position. It is important to note that this device is not capable of determining absolute position unless some external method is provided for locating a datum. Also, in its simplest form, the incremental encoder—with only one row of slits—is unable to determine the direction of rotation. The double offset rows shown in Figure 8.40 allow the electronics to determine direction based on the relative occurrence time of the pulse from each row. The resolution of encoders is determined by the number and spacing of the slits.

An *absolute encoder* operates on the same principle as the incremental encoder but uses additional tracks, or rows of slits, to provide an absolute position indication. Figure 8.41a shows a segment of such a disk. Here, five tracks are used. Four of these provide the absolute position information, and the fifth one—called the *anti-ambiguity track*—is used as an internal self-check by the encoder. The pulse pattern produced from this disk is represented in Figure 8.41b. This pattern is converted to a binary code that represents the absolute position of the rotating element with reference to a known datum. This datum is coded on the disk, so no external reference is needed.

Again, the resolution depends on the number and spacing of the slits; however, since absolute position is the required output, greater accuracy can be achieved by adding tracks, usually in multiples of four. The pulse pattern shown in Figure 8.41b results in a 4-bit binary code. Multiples of four provide the common 8-, 12-, and 16-bit binary codes used by most computers. In reality, the binary output code from these. units is prone to giving false readings at the instant of an indicated position change, particularly when all the binary digits change to indicate the new position. For example, in going from position 0111 to position 1000, it is unlikely that all digits will change at precisely the same time. As a result, we may see a transition such as 0111—0110—0010—1010—1000, or something similar. Each of these transitions would indicate a specific (and incorrect) position. To

Figure 8.41 An absolute encoder uses five rows of slits to provide rotary position information. (a) Encoder disk. (b) Pulse pattern. (Source: Intelligent Sensors, edited by Ryoji Ohba. © 1992 by John Wiley & Sons Limited. Reproduced by permission of John Wiley & Sons Limited)

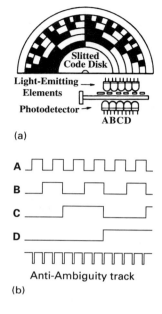

(a)

(b)

overcome this problem, a *Gray code* disk can be used. In Gray code, only one bit changes at a time, so the transition problem is eliminated. A 4-bit Gray code is as follows:

Decimal	Gray code
0	0000
1	0001
2	0011
3	0010
4	0110
5	0111
6	0101
7	0100
8	1100
9	1101
10	1111
11	1110
12	1010
13	1011
14	1001
15	1000

A different type of encoder uses an etched copper disk with carbon or copper brushes that ride over the individual tracks. These devices are less expensive than the light units and can work over a wide range of signal levels. They are less robust than the light units and have a relatively short life due to brush and track wear.

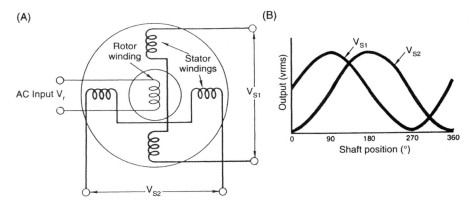

Figure 8.42 A resolver is basically a rotary transformer. The shaft angle is calculated from the sinusoidal output signals. (Courtesy of *Sensors Magazine* (www.sensorsmag.com), Peterborough, N.H.)

A *resolver* is an analog induction device—essentially a rotary transformer. It consists of one rotating winding and two stator windings. As illustrated in Figure 8.42, the stator winding are set at 90° to one another. The rotor winding is excited by an AC signal—usually 400 Hz or higher. The outputs induced from the two stator windings are proportional to the sine and cosine, respectively, of the angle between them and the rotor. When attached to a shaft, these output signals can be resolved by the computer to indicate the precise angular position of the shaft. In some designs the excitation signal is applied to the stator windings, with the output signal obtained from the rotor. In this case, the stator signals have a 90° phase shift.

Like resolvers, *synchros* are analog induction devices. They typically consist of three stator windings set at 120° and a single rotor winding, as shown in Figure 8.43. Again, an AC supply (usually 400 Hz or 50 Hz) excites the rotor winding, which results in an induced voltage in the three stator windings. The magnitude and phase of these individual voltages uniquely define the position of the rotor.

A *rotary variable differential transformer* (RVDT) uses a specially shaped ferromagnetic rotor that simulates the linear displacement of the core in an LVDT. These units are capable of continuous rotation but are normally used within an operating range of ±40°.

8.6.6 Velocity and Acceleration Measurement

Velocity and acceleration can be obtained from most of the linear and rotary position transducers already discussed by including rate-of-change calculations in the computer software.

There are numerous other techniques for determining angular velocity. These normally involve digital devices that produce a pulsed output. Typical of these are magnetic pickups that sense the passage of "teeth" on a rotating disk, and electro-optical units that detect pulses of light through holes or slits in a rotating disk (similar to encoders). In addition, photoreflective devices can be used to detect the passing of a strip of reflective tape on a rotating shaft.

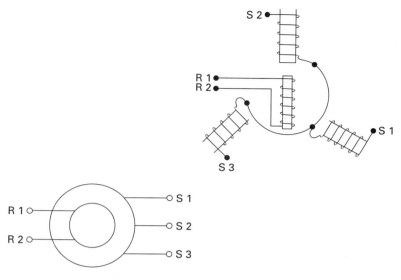

Figure 8.43 A synchro is similar in concept to a resolver but uses only three stator windings.

Analog measurement of rotational speed is provided by a tachometer-generator, usu-ally called a *tachgenerator*. This device is basically a generator whose output is essen-tially linearly proportional to its rotational speed up to its maximum, or saturated, speed. Both AC and DC output tachgenerators are available.

8.7   SUMMARY

Transducers and sensors are essential for feedback and control of automated machines. They vary in function, design, capability, accuracy, and precision and must be selected to satisfy each specific application.

Sensors and transducers are characterized as discrete or continuous depending on whether they sense a specific event or a continuous activity. The "family tree" shown be-low lists each type.

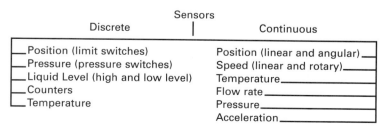

Many discrete sensors can be used as switches to directly control the motion of the machine or a process. They may also be used as inputs to a controller such as a programmable logic controller. In their most simple application, they may be used to turn on a light or sound an alarm.

Continuous sensors, like some discrete sensors, may be used merely to gather and present information. A pressure transducer, for example, may be used to provide an input for a digital pressure gauge. Continuous sensors also may provide continuous feedback and control information to programmable logic controllers or the operational amplifiers in servo controllers.

Usually, the information signal from a transducer is analog. If it is to be used by a computer, it is almost always necessary to convert the signal to a digital format using an analog-to-digital converter.

SUGGESTED ADDITIONAL READING

Handbook of Photoelectric Sensing, 2d ed. 1993. Minneapolis, Minn.: Banner Engineering Corp.

Ohba, Ryoji. 1992. *Intelligent Sensor Technology.* New York: Wiley.

Pressure Transducer Handbook. Pasadena, Calif.: CEC Instruments.

Parr, E. A. *Industrial Control Handbook.* vol. 1: *Transducers.* 1987. New York: Industrial Press Inc.

REFERENCES

Balluff, Inc. 1987. *Electromechanical Single and Double Limit Switches with Accessories.* Catalogue No. 103E, Issue #705. Florence, Ky.

Balluff, Inc. 1988. *Non-Contact Switching.* Catalogue No. 321E, Edition 8806. Florence, Ky.

REVIEW QUESTIONS

General

1. Explain the difference between a transducer and a sensor.
2. Explain the difference between analog and digital signals.
3. Define *resolution.*
4. Explain the difference between accuracy and precision.
5. Explain the difference between repeatability and reproducibility.
6. Explain the difference between discrete and continuous sensing.
7. Explain the difference between inductive and capacitive proximity switches and their applications.

8. Explain the operation of reed and Hall effect switches and discuss their differences.
9. In the context of photoelectric sensors, explain the following terms:
   a. Opposed mode
   b. Retroreflective mode
   c. Diffuse mode
10. Explain the terms *light operate* and *dark operate.*
11. Which type of pressure gauge is best for sensing very small pressure changes?
12. Which type of transducer has the highest frequency response?
13. Explain the difference between absolute and incremental encoders.

CHAPTER 9

Programmable Logic Controllers

OBJECTIVES

When you have finished this chapter, you will be able to:

- Describe the architecture of a programmable controller.
- Discuss the different types of memory used in programmable controllers.
- Explain the concepts involved in connecting input and output devices to the terminals of programmable controllers.
- Explain the difference between dedicated and general-purpose programming devices.
- Draw, read, and interpret ladder logic diagrams.
- Write simple PLC programs using Boolean mnemonics.

9.1   INTRODUCTION

A programmable logic controller (PLC) is a device used to provide automatic, and easily changed, control for an almost endless variety of motion systems. In Figure 9.1 we see a simple example of a PLC application. In this circuit we have three hydraulic cylinders that must be operated in a specific sequence. The sequential operation is controlled by the PLC. To initiate the operation, the two START push buttons must be pushed and held. This input action causes the PLC to produce an output signal to energize solenoid A, causing cylinder A to extend. When cylinder A has extended a prescribed distance, it contacts limit switch A. This limit switch provides an input to the PLC to trigger an output to deenergize solenoid A and energize solenoid B. The sequence repeats for solenoid C. When cylinder C contacts limit switch C, that limit

Figure 9.1 Three-axis electrohydraulic can crusher.

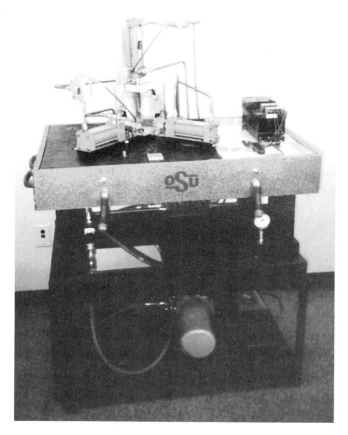

switch provides an input to the PLC to energize all three opposite solenoids and cause the cylinders to retract.

This operation seems to be nothing special. In fact, it is the same can crusher circuit we saw in Chapter 4. Although we do not need a PLC to do the job, it does provide flexibility for us. Suppose, for example, that we decided to change the sequence to operate cylinder C before cylinder B. Without the PLC, we would have to physically rewire the system. With the PLC, we simply rewrite the program. In this simple system there really is not much effort involved in the rewiring, but if the PLC was controlling tens or even hundreds of outputs, the job could be daunting. Reprogramming would be much easier, faster, and less expensive. As an additional benefit, the PLC usually eliminates the need for the relays that were used in previous circuits.

From this simple example we have seen that a PLC receives a series of inputs from a variety of switches and transducers, manipulates them according to a program that is written by the user, and provides a series of outputs that control the sequence of operation of the output devices. This sequence is shown as a block diagram in Figure

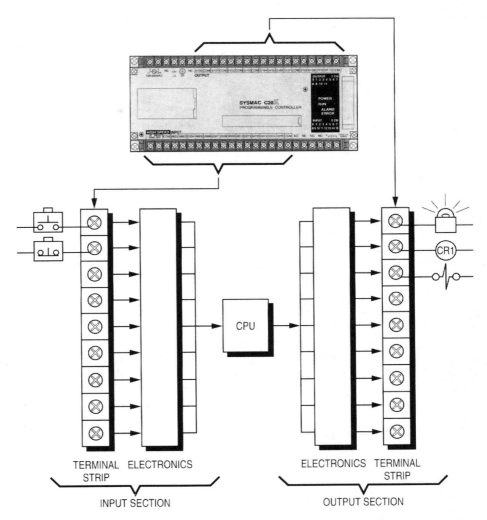

Figure 9.2 This block diagram shows the input and output arrangements for a typical small PLC. (Courtesy of Womack Machine Supply Company)

9.2. Physically, the input devices are completely independent of one another, as are the output devices. The logic that connects them into a system is contained in the program in the PLC.

In this chapter we take a general look at programmable logic controllers as they are applied in industry. We start with some background information, followed by a brief look at the various components that make up a programmable controller. We then go into programming languages, ladder logic diagrams, and some examples of PLC applications. We will also look closely at two different programming schemes—the use of dedicated programming terminals and the personal computer.

9.2   BACKGROUND

A programmable logic controller is a dedicated electronic device (actually, a small computer) used to provide the control logic for machinery, equipment, or complete processes. In fact, an entire manufacturing plant can be controlled by a single master PLC, which may be controlling and receiving information from a number of satellite controllers and input devices. Because of the logic functions and the industrial applications, these devices are also known as *programmable controllers* (PC) and *programmable industrial controllers* (PIC). In this text we will use the PLC designation. Most PLCs are specially constructed and packaged to "harden" them for the rigors of their somewhat hostile environments.

The concept of the programmable controller originated in the automotive industry. In the process of changing manufacturing facilities to produce new car models, very extensive and expensive relay control systems had to be changed. It was easier to start from scratch than to rewire the control circuitry; therefore, huge relay panels were simply torn out and scrapped. This was obviously a very expensive procedure. Even though it was quicker than rewiring, it was still costly and very time consuming, leading to long and expensive downtimes.

In 1966 the Hydramatic Division of General Motors generated the first specifications for a programmable controller. In 1969 Gould (then Modicon) delivered the first production PLCs to General Motors. The concept met with instant acceptance, and in 1971 the first PLCs outside the automotive industry went into operation.

These initial units were little more than very sophisticated switches, but as with most electronic devices, the capabilities and sophistication of PLCs grew very rapidly. In 1973 so-called smart PLCs became available, which had the ability to compare input and output (I/O) signals (feedback), interface with printers, CRTs, and other devices. By 1975, analog PID (proportional integrating derivative) PLCs could interface with analog input devices such as thermocouples and pressure transducers. (Previous devices could respond only to off-on signals such as limit switches and optical sensors.)

Since then the applications of PLCs have expanded to include hierarchial PLC systems for integrated manufacturing systems, and integration of entire plant operations through PLC communication systems. Improvements in input and output modules have brought about higher speeds, shorter response times, and improved control. Improvements in programming devices and languages have greatly simplified operator interfacing. A typical small PLC is shown in Figure 9.3.

Figure 9.3   A typical small PLC with a dedicated programming module. (Courtesy of Womack Machine Supply Company)

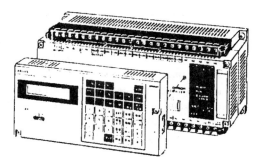

9.3   PLC ARCHITECTURE

All programmable controllers have certain basic components, although the form of these components may differ significantly. Figure 9.4 illustrates these basic components.

9.3.1 Central Processing Unit

The central processing unit (or CPU) is the brain of the PLC. It is, at least in most modern PLCs, a microprocessor-based unit that contains all the logic and control algorithms for the controller. The CPU scans the total information package (memory and I/O devices) continuously. The time required for this scan depends on the configuration of the system, the number of I/O units, the number of special instructions, and the number of peripheral devices being used as well as the speed of the microprocessor. A typical scan will include the following operations:

- Check the memory and the I/O bus.
- Read input data from input devices.
- Execute instructions based on input data.
- Send appropriate output responses to the output devices.
- Service peripheral devices (update data acquisition systems, show condition changes on monitors, etc.).

Most PLC manufacturers (especially of the larger units) list the scan times of their units "per K of memory." (This refers to 1024 words of memory.) Depending on the microprocessor used, this scan time varies from about 0.4 ms to as much as 60 ms per K of memory. You will generally find that the smaller PLCs have longer scan times than the large units. Because the small units have less memory, fewer I/O devices, and less capacity for special instructions and peripheral equipment, they can have a slower scan rate and still have fully acceptable speed.

9.3.2 Memory

The memory stores digital control logic, the process program, and instructions required to operate the system. These data are stored in the memory in units called *bytes*. A byte

Figure 9.4   PLC architecture.

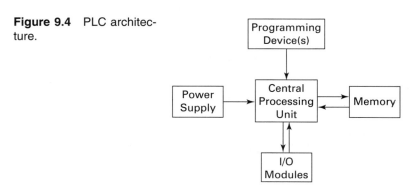

is a grouping of eight *bits*, which are single binary digits (0 or 1). A *word* of memory will consist of one to three bytes, depending on the capability and design of the memory unit.

The memory may lose its information when the power is turned off or it may retain it. A memory that retains its information is termed *nonvolatile*, whereas a *volatile* memory loses its information. The user interface with the PLC usually involves volatile memories, because they are the easier of the two to program and subsequently erase and reprogram; however, their volatile characteristic requires that a backup battery be provided to retain the memory when the power is off.

The digital memory used in modern microprocessors may be classified into four categories:

- RAM—Random-Access Memory
- ROM—Read Only Memory
- PROM—Programmable Read-Only Memory
- EPROM—Erasable Programmable Read-Only Memory

Random-access memory (RAM) is the only volatile memory of the four. More than 90% of microprocessor-based PLCs use RAM (Childs 1987). RAM is preferred because it is easy to program and reprogram. It provides massive memory capability to store programs and data that can be accessed quickly. Any memory element can be accessed at any time and in any order. RAM is made up of either flip-flops (bistable digital elements) or latches, which are combinational electronic circuits. Programming a RAM simply employs binary logic to set the memory element to a 0 or 1 condition. A variation on the RAM is dynamic RAM, or DRAM. RAMs are preferred over DRAMs because larger capacity and higher density arrays are possible with RAMs; however, RAMs require more physical space and are more expensive than DRAMs (Stubbins 1986).

Read-only memories (ROMs) are memory cells that store information permanently. The most common use of ROMs is to store the operating program of a computer so it will run properly on startup. They may also be used for permanent, unalterable process control. In this application they are usually provided with the equipment as it is delivered from the manufacturer. Many of the computers in automobile applications are ROMs, including the synthesized voice warning systems. ROMs are fabricated by omitting specified digital cells to provide the required bit pattern. Because this is actually a physical presence or absence of a physical component, once programmed (manufactured, really), the ROM cannot be altered.

Programmable ROMs (PROMs) are manufactured with a full grid (or array) of memory cells, each of which is electronically tied to a signal grid by a fusible link. Programming a PROM involves physically melting the fusible link to disable specified memory cells. The result is a pattern that, as in a ROM, cannot be altered. Once the link has been melted, the cell cannot be restored.

Erasable PROMs (EPROMs) are often more convenient than either ROM or PROM, because they can be erased and reprogrammed. Unlike programming the previous devices, programming an EPROM involves the electrical, rather than physical, isolation of the memory storage cells. Subjecting the EPROM to intense ultraviolet light for about 30 min restores the cell connections by causing the isolating electrical charge to leak off. The entire memory array can then be reprogrammed by reintroducing isolating charges in a different pattern.

Electrically alterable read-only memories (EAROM) and electrically erasable programmable read-only memories (EEPROM) are also in the EPROM family. They differ from EPROMs in that an electrical signal, rather than ultraviolet light, is used to change the original memory cell pattern. All these read-only-type devices are nonvolatile and do not require power to retain their programs.

9.3.3 Input/Output Modules

The input/output (I/O) modules provide the physical and electronic interfacing between the CPU and the devices that provide and receive electrical signals. Input devices include digital and analog units. Digital devices include toggle switches, push buttons, limit switches, pressure switches, and others that provide discrete off-on signals. For process control, touch screens are widely used. These screens allow the operator to turn devices off and on, control liquid levels and feed rates, and perform numerous other control functions simply by touching the appropriate symbol on the screen. Analog devices such as pressure sensors, thermocouples, and linear variable differential transformers (LVDT) provide continuous, varying signals that represent the state of the system. If the PLC does not have PID circuitry, it is usually necessary to provide an external unit to convert the analog signal to a digital signal (A/D converter).

The input module will probably be assigned some signal conditioning functions to ensure that a usable signal is presented to the CPU. It will almost certainly provide circuit protection capabilities to prevent damage to the CPU circuitry in case of faults, surges, or spikes in the incoming signal. This is often accomplished through optical isolation at the interface to the CPU.

The output module provides the circuitry for transmitting signals from the CPU to lights, solenoids, motor starters, signal devices, status monitors, and other devices. These signals are normally digital. Therefore, the output module really functions as a switching device, providing on-off signals to its associated devices. In some cases the output module is not able to handle the power required by the output device, so it sends the appropriate signal to an external relay. This relay, in turn, assumes the appropriate state, and its contacts handle the actual power circuit.

The input and output devices are external to the PLC. They are physically connected to the hardware by wires connected to the terminal strips on the PLC. Figure 9.5 shows a typical arrangement. There is no physical connection between the inputs and outputs.

Each I/O terminal has a unique address. In this example, we have used a simple system. Inputs are 0000 through 0010, and outputs are 0100 through 0102. Here, the first two digits designate the specific function of the address (00 for inputs, 01 for outputs) as well as their physical location on the PLC. Many address schemes are used, depending on the specific PLC. When programming the operation, it is critical to address each I/O point properly.

If solid-state devices are to be used to provide either input or output signals for DC circuits, it is critical that the correct type of device be selected. Solid-state devices can be simplistically described as switches that have no moving parts such as those found in mechanical switches. For DC circuitry, these solid-state switches are termed *transistors*.

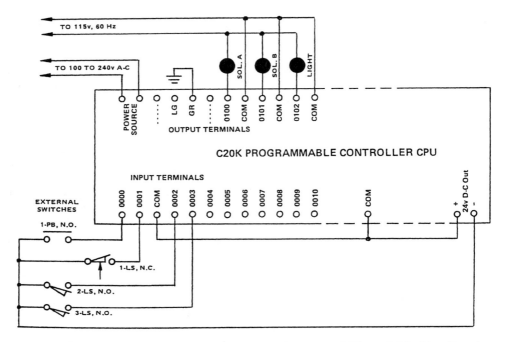

Figure 9.5 Connections to controller terminals. (Courtesy of Womack Machine Supply Company)

A transistor is a sandwich made up of thin slices of semiconductor materials, usually silicon and germanium. From an activation point of view, there are two types of transistors—NPN and PNP. These designations describe the polarity of the individual slices of the sandwich, as shown in Figure 9.6. In an NPN device, the two outer slices are negatively polarized, while the middle slice has a positive polarity. In the PNP device, the outer slices are positive, while the middle slice is negative. To actuate the transistor and complete the circuit, a signal of the same polarity must be applied to the middle slice.

In somewhat more technical terms, the three slices are designated as the base (B), collector (C), and emitter (E), as shown in Figure 9.7. The connection of these devices is critical. The control signal is always connected to the base lead, but it must be either positive or negative depending on the type of transistor used, which, in turn, depends on the characteristics of the PLC to which the device is connected.

In Figure 9.7a, we see the symbol for an NPN transistor with the base, emitter, and collector connections. Notice the arrow between the base and the emitter. This indicates that this is where the (electronic) switching takes place. In this case, the circuit is complete through C to B; however, the circuit from B to E (hence to ground or negative) is completed only when there is a positive control signal applied at B. Notice, also, that the load is placed *between the positive terminal of the power supply and C.*

The symbol for a PNP element is shown in Figure 9.7b. Here, the switching takes place between E and B (note the direction of the arrow). The circuit is completed only

Figure 9.6 A transistor consists of three slices of a semiconductor with the middle slice having an electrical charge opposite that of the end slices.

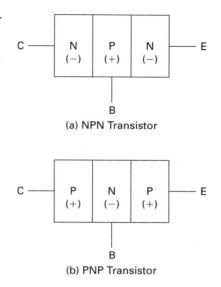

(a) NPN Transistor

(b) PNP Transistor

when a negative signal is applied at B. In this case, the load is connected *between the negative terminal of the power supply and E.*

In selecting the type of solid-state switching device to use with a specific PLC, you must first determine whether the I/O modules are AC or DC. If they are AC, a *triac* is used (so this discussion does not apply); however, if they are DC, then you must determine whether to use PNP or NPN. The PLC I/Os will be designated as *current sourcing* or *current sinking*. These two terms convey the same meaning (electronically) as we have just discussed. Sourcing modules require PNP elements, whereas sinking modules require NPN elements.

The type of PLC module also dictates the way the load must be connected to the output terminals of the PLC. If the module is current sinking, the *negative* terminal of the DC power supply must be connected to the common terminal on the PLC. The positive

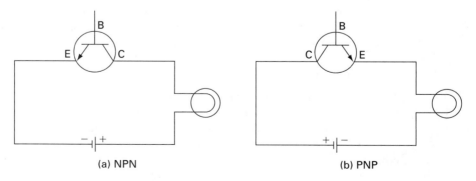

(a) NPN   (b) PNP

Figure 9.7 NPN and PNP circuit connections.

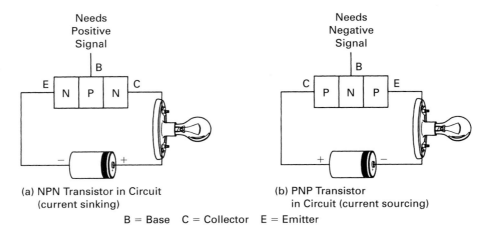

Figure 9.8 Transistors used in sinking and sourcing circuits. (Courtesy of Womack Educational Publications)

terminal of the power supply must go to the load, which is in turn connected to the designated output terminal of the PLC, as shown in Figure 9.8a.

The connections for a sourcing module are shown in Figure 9.8b. Here, the *positive* terminal of the DC power supply is connected to the common terminal of the PLC, and the negative terminal connects to the load. The load then connects to the designated output terminal on the PLC.

Most equipment manufactured in the United States and Japan uses NPN transistors. European manufacturers favor PNP devices.

In most PLCs an internal power supply provides power for the internal operation of the unit and a minimal current for the input devices. Output devices, however, frequently require an external power supply. This device must be matched to the parameters of the output module.

The I/O modules in most PLCs are removable, replaceable modules. Therefore, you can choose the type module to meet your particular needs (24 VDC, 120 VAC, etc). Be careful of the power limitations as you choose (and use) output modules, especially.

There is a huge range of sizes of PLCs available. They may be very small and include as few as eight inputs and outputs, or they may be physically quite large and include several thousand I/Os.

9.3.4 Programming Devices

The programming device provides the interface between the programmer (user) and the CPU. They can take on many different forms but can be divided into those that are dedicated to the particular PLC and those that are general purpose (usually, a personal computer). A general-purpose programmer will usually require software peculiar to the specific PLC.

All programming devices have a keyboard of some type. Dedicated programmers often use a small, abbreviated keyboard containing relatively few keys, each of which

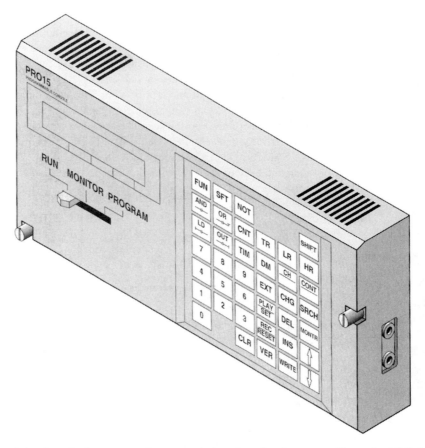

Figure 9.9 A typical programming console has function keys specific to the PLC for which it is intended. (Courtesy of Womack Educational Publications)

triggers a complete command or functional element when touched. One such device is shown in Figure 9.9.

Programming devices normally include some sort of visual display to assist in entering and editing programs. On dedicated terminals there may be a small LED display along with function indicator lights. Other dedicated units might include a full CRT display complete with superior graphics. Personal computers normally utilize their associated monitors.

The programming device may program the PLC directly, or it may simply facilitate writing the program that will subsequently be downloaded to a programming tape, EPROM, or other such device for eventual uploading to the CPU of the PLC. In the case of a personal computer it is often possible to communicate with the PLC through some sort of interfacing module. This is frequently done in laboratory settings and when initiating a new process. In actual application, however, the program is usually loaded into the PLC so that it operates independently of the personal computer.

Later in this chapter we will look at some simple programming applications showing both types of programming. To facilitate your understanding of these processes, we will use commonly available, real-world software and hardware for our examples. For the general-purpose programmer, we will use the Automation Studio software from FAMIC Technologies 2000. With this software you can create the hydraulic circuit, the relay ladder diagram, and the ladder logic diagram on your computer monitor. Then, using the included simulation capability, you can simulate the operation of both the electrical control circuit and the hydraulic circuit on the screen. This allows you to create, modify, and simulate the operation of the entire system before you begin to assemble the hardware or program the PLC. FAMIC also produces interface cards that allow you to use your computer to control the actual system or to download your completed ladder logic diagram into the PLC in the correct language and syntax. Included with the instructor's manual for this text is an Automation Studio demonstration disk that will allow you to draw and simulate some simple circuits.

For the dedicated programmer examples we will use the OMRON SYSMAC C20K programmable controller. The C20K is a shoebox-size PLC with 20 I/O. It is programmed in three ways. The examples we will discuss here will use Boolean mnemonic code input through the dedicated keypad (which OMRON terms the programming console). Software is available to allow a ladder logic diagram to be downloaded directly to the PLC, saving you the effort of converting your ladder logic into Boolean mnemonic code and then "punching it in." The third method is to create an EPROM that can be inserted into an EPROM socket on the PLC. This can be very convenient when numerous processes are operated individually by one PLC. Changing the program is as simple as unplugging one EPROM and inserting another.

9.3.5 PLC Sizes

Programmable controllers, with some very broad generalizations, can be divided into five groupings, as shown in Table 9.1. These groupings are based on the number of I/Os,

Table 9.1 PLC Size Groupings

Size	No. of I/Os	General Applications	Math Capability*
Mini	Up to 32	Relay replacement	Little
Micro	33–64	Relay replacement, timers, counters	Some
Small	65–128	Same as Micro; used for material handling control and some process control	Yes
Medium	129–892	All of Small plus data collection	Extensive
Large	892+	All of Medium; may also control a number of smaller PLCs and work cells as well as generate reports	Extensive

*Math capability varies widely. Some large PLCs have no math capability at all, whereas some micros have all arithmetic and some trig functions.

memory size, and functional capabilities. Other sources omit the mini category and size strictly according to the number of I/Os. There are problems with any classification system because of the vast differences among "similar" PLCs. For the "large" category, be aware that 892 I/Os is the *minimum* for this ranking. Some PLCs in this grouping have in excess of 40,000 discrete I/O points.

9.4   PROGRAMMING LANGUAGES

Many different programming languages are used to input instructions to PLCs. According to Childs (1987), these can be grouped into four major categories:

- Ladder diagrams
- Boolean mnemonics
- Functional blocks
- Plain-language statements

Functional blocks and plain-language statements (sometimes termed "literal" programming) are considered to be high-level languages. Included among these high-level languages are FORTRAN, BASIC, PASCAL, PML, C, and the like, as well as several proprietary languages developed for specific applications.

Ladder diagrams and Boolean mnemonics are considered to be the basic-level languages (not to be confused with the BASIC programming language). Boolean mnemonics are so named because of their derivation from Boolean algebra. They involve the use of AND, OR, NOT, and similar functions in describing input and output conditions.

Ladder diagrams are older than PLCs themselves. For years before the first PLCs were put on line in 1969, electricians had been working with ladder diagrams for relay circuits, because they could easily depict the relay logic. At that time, relatively few people outside the computer industry, academia, and some scientific areas had any knowledge of computer programming. These two circumstances led to the development of ladder logic, which is a logical evolution and adaptation of relay logic ladder diagrams. This provided a system that was readily accepted by both the computer scientists and the people who were to use and maintain the equipment. The similarity to relay logic dispelled much of the mystique of these new high-tech devices and made them less intimidating to the electricians and maintenance personnel.

Although the use of Boolean mnemonics and the high-level languages is more efficient from a pure computer science viewpoint, ladder logic is still the most popular of all the programming methods because of its practical simplicity and its ease of application by those not well versed in computer languages. In many cases it provides the ability to simulate the operation before actually connecting the system.

Although ladder logic programming is a relatively simple method for electricians and others who work with PLCs, it still seems cumbersome and difficult to those outside these fields. In an effort to develop a more universal and easily communicated system, a design team consisting of academic and industrial representatives was formed in 1979 by the French Association for Economical and Applied Cybernetics. This team developed a graphical method for specifying industrial automation that uses simple syntax, graphical representation, and concise commands. The method was given the name GRAFCET (an acronym for *GRAphe Fonctionnel de Commande Etape/Transition* or, in English, Step Transition Function Charts) (Baracos 1992).

GRAFCET offers two major advantages to those involved in automation processes. First, there is no need to understand ladder logic concepts. Second, there is no need to be familiar with the programming of any PLC. Although the GRAFCET diagrams can easily be translated into ladder logic, software is available that will translate the GRAFCET flowcharts directly into programs for most major PLC types. In most cases, the compiler software will transfer the program directly from a computer to the PLC memory. This functionally allows a person who is PLC illiterate to program any PLC. It also facilitates the changeover from one type of PLC to another that might occur during plant modernization or process changes. These capabilities have led to worldwide acceptance of GRAFCET as a tool for industrial automation. It is particularly applicable to electrohydraulic and electropneumatic applications where sequential operation is the norm.

9.5   THE TRANSITION: RELAY LOGIC TO PLC LOGIC

In Chapter 3 we learned to express an electrical wiring diagram as a ladder diagram that utilizes standard symbols to represent various switching elements (inputs) and output devices. The logic of these ladder diagrams is easily understood, because the electrical flow can readily be envisioned. Each rung of logic represents an electrical circuit, and the operation of each circuit is clear—if the circuit is completed through any combination of switching elements, the output will result. If the circuit is not completed, there is no output.

With this understanding, the transition to PLC logic ladders (we will just call it *ladder logic* from here) will be a simple step. It is simple because the logic elements used for all types of switching elements are the same normally open and normally closed contact symbols we have already used. The output devices are often simply appropriately labeled circles (or sometimes rectangles).

Let's look at an example. In Figure 3.15 we used a toggle switch to start an electric motor. Figure 9.10 repeats that circuit and then translates it into ladder logic. Additionally, that ladder logic diagram represents the push-button circuit shown in Figure 3.18.

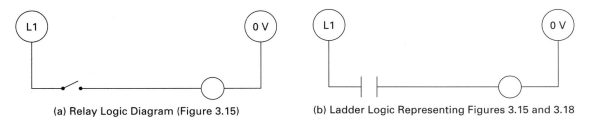

(a) Relay Logic Diagram (Figure 3.15)   (b) Ladder Logic Representing Figures 3.15 and 3.18

Figure 9.10   Logic diagrams for Figures 3.15 and 3.18.

Figure 9.11 Ladder logic diagram for Figures 3.13 and 3.21.

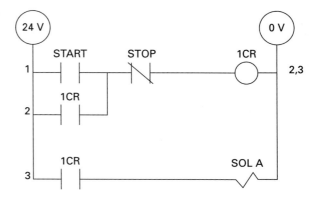

Figure 3.21, which already represented a considerable simplification of the wiring diagram of Figure 3.13, can be further reduced now to the ladder logic shown in Figure 9.11.

Notice that in these ladder logic diagrams we adhere to the same rules that were introduced in Chapter 3. Specifically, place the hot line on the top (horizontal ladder) or left (vertical ladder); work from hot to ground or neutral with all switching elements on the hot side of the output; label all logic elements appropriately. There is no absolute tradition for the label format. For instance, you will see relays labeled 1CR and 1-CR. They may even be called Relay 1 or Rly 1—whatever is convenient. (Just be sure to use the same label on all their contacts. This also applies to the contacts associated with timers and counters.)

If you understand these examples, you have made the transition and are ready to do the practice exercises at the end of the chapter.

9.6   APPLICATION EXAMPLES

In this section we will take a look at some examples of fluid power circuits that are controlled by a programmable controller. We will examine the schematic of the fluid power circuit as well as the relay ladder diagram. We will then translate the relay ladder into a ladder logic diagram as it might appear on the monitor display of a general-purpose programmer, and into Boolean mnemonics for entry into the programming keyboard of a commonly used PLC.

Figure 9.12 shows a simple fluid power circuit that is used to start and stop a hydraulic motor. The relay ladder control circuit, shown in Figure 9.13, consists of START and STOP push buttons and a relay with a holding circuit. Figure 9.14 shows the ladder logic diagram for the circuit and lists the Boolean mnemonic program that would be used to program the OMRON C20K. In the ladder diagram shown in Figure 9.13 all the control devices (the push buttons and the relay—perhaps even the relay contacts) are hardware that you can see and touch. In the *ladder logic* diagram of Figure 9.14, however, only the START and STOP (designated 0003 and 0004, respectively) push buttons are external to the PLC. The relay (1000) and contacts are internal. Even the output (0103)

Figure 9.12 Hydraulic motor circuit.

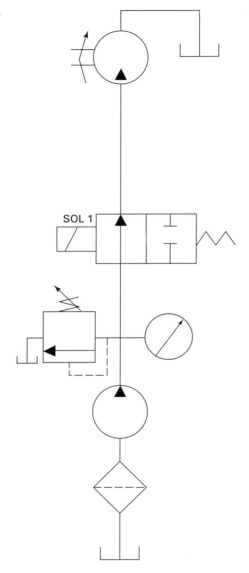

represents an internal switch rather than the actual solenoid. The solenoid itself, along with its power supply (if needed), would have to be wired to the specified output terminal. The PLC program in Boolean mnemonics is specific to the OMRON family of PLCs.

Looking back at some circuits used in Chapter 4 can provide some useful ideas, also. For instance, Figure 4.39 (repeated here as Figure 9.15) is a circuit that provides for a single-cycle automatic reciprocation of a hydraulic cylinder. The associated electrical ladder diagram is also shown. Figure 9.16 shows the ladder logic diagram and the Boolean mnemonics program to control this circuit. The START pushbutton is an external input

Figure 9.13 Electrical ladder diagram for Figure 9.12.

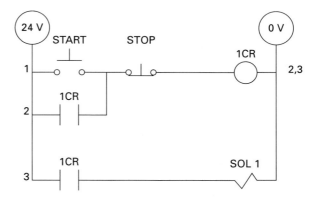

and is designated 0003. The external limit switch is represented by the normally closed contacts 0004. The output to the solenoid is shown as 0103. The coil labeled 1000 and its contacts are internal.

The timer circuit of Figure 4.40 (repeated here as Figure 9.17; see page 250) can be programmed as in Figure 9.18 (see page 251). The external limit switch (LS1) is represented by 0005. Again, the external inputs (push buttons) are labeled 0003 and 0004. Coil 1000, timer TIM01, and their contacts are internal.

Figure 9.19 (see page 251) may help to clear up any confusion you are feeling right now. This figure illustrates the PLC connections that could be used to operate the circuit of Figure 9.18. On the left-hand side of the PLC are all the external input devices—a normally open push button (START), a normally closed push button (STOP), and a normally open limit switch (LS1). These are shown connected to PLC input terminals 0003, 0004, and 0005, respectively. Note that the terminal numbers coincide with the addresses in the program. When the PLC mode switch is placed in the RUN position, it will begin to scan the designated addresses to determine the state of each input.

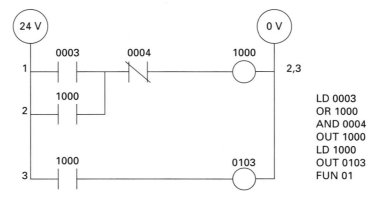

Figure 9.14 Ladder logic diagram and Boolean mnemonics program for the ladder in Figure 9.13.

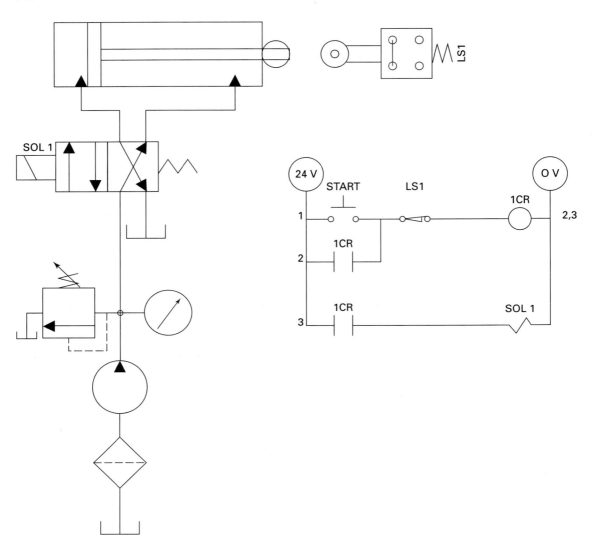

Figure 9.15 Hydraulic and electrical ladder diagrams for a one-time automatic reciprocation of a hydraulic cylinder.

The output device (the solenoid) is shown connected to the 0102 output terminal, as designated in the program. Because this C20K uses a current-sinking output, the negative terminal of the power supply (if needed) would be connected to the common terminal on the output panel.

Everything else in the ladder logic diagram (the relay (1000), its associated contacts, the timer (TIM 01), and its associated contacts) are inside the PLC. When the computer finds the correct logic in rungs 1 and 2, it completes the internal circuit between

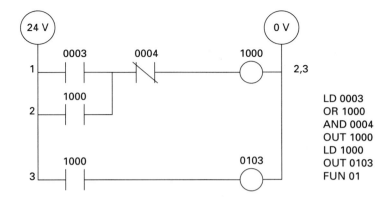

Figure 9.16 Ladder logic and Boolean mnemonics program for Figure 9.15.

output terminal 0102 and its common (functionally, it closes the switch between them), allowing solenoid A to be energized. Subsequently, the cylinder extends and contacts the (external) limit switch. This completes the logic to start the (internal) timer, which ultimately opens the (internal) timer contacts, disrupts the logic for the (internal) relay, and opens the (internal) contacts for output 0102. This deenergizes the (external) solenoid, allows the valve to shift, and causes the cylinder to retract.

Now let's look at a simple example of the flexibility of the PLC and see why it has found such widespread use. Suppose we had a control circuit that used two normally open pushbuttons and one normally closed pushbutton to control a light. We could connect the pushbuttons to the PLC terminal strip as shown in Figure 9.20. If we wanted START 1 and START 2 to be in series, we would program the PLC using AND logic. This would tell the computer to look for inputs on *both* terminals 0002 AND 0003 as well as 0004. The Boolean mnemonic program for the OMRON C20K would be as shown in Figure 9.21.

Now suppose we change our minds and decide that START 1 and START 2 should be in parallel. This involves OR logic. We make this change simply by changing one line in the program as shown in Figure 9.21b. Now the computer looks for an input on *either* 0002 or 0003 (or both) AND 0004. No wiring changes are necessary. Once the input devices have been connected to the terminal strip, the way in which they are used is purely a function of the program.

The exercises at the end of this chapter will give you the opportunity to explore other control circuits and their PLC programs. Remember that these are rather simplistic systems, so the value of the PLC in these applications is not extremely impressive. Imagine, though, a complex process control circuit including hundreds of hardwired relays. A process change would involve physically rewiring the entire relay panel. If the process was controlled by a PLC, a process change would involve only reprogramming the PLC. Although this may not be a simple task, the program can be written (and often programmed and tested through simulators or emulators) before the change is to be made. At that point it may be a simple matter of downloading the program to the PLC.

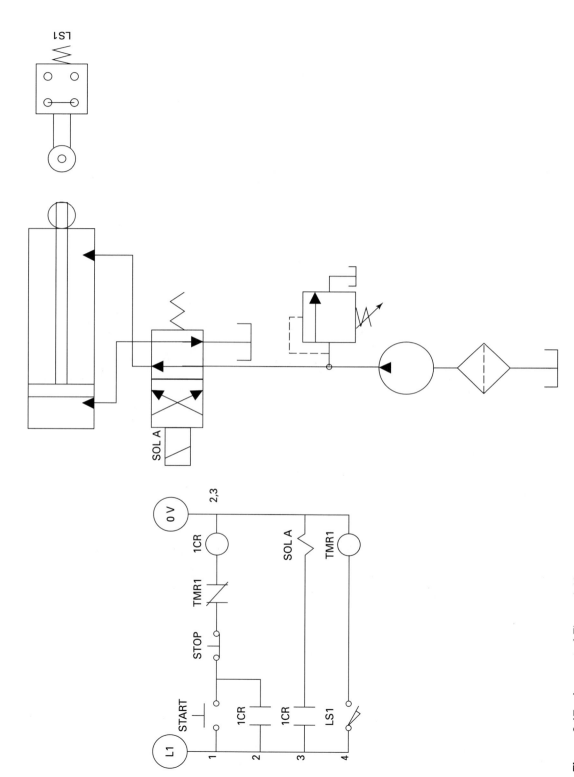

Figure 9.17 A repeat of Figure 4.40.

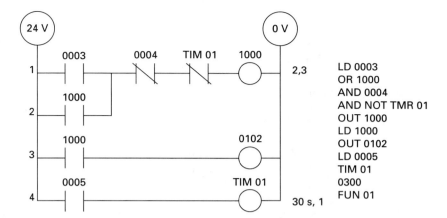

Figure 9.18 Ladder logic and Boolean mnemonics program for Figure 9.17.

LD 0003
OR 1000
AND 0004
AND NOT TMR 01
OUT 1000
LD 1000
OUT 0102
LD 0005
TIM 01
0300
FUN 01

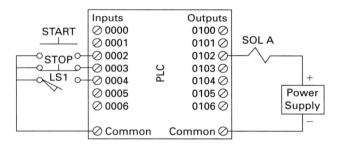

Figure 9.19 External PLC connections for the circuit of Figure 9.17.

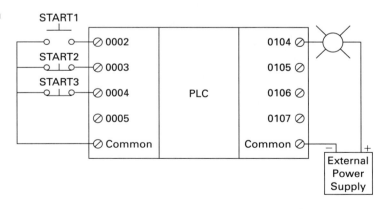

Figure 9.20 Three push buttons control one light.

Figure 9.21 These are two of the logic options for controlling the light in Figure 9.20. The wiring remains the same, but the logic is different.

LD 0002
AND 0003
AND 0004
OUT 0104
END

LD 0002
OR 0003
AND 0004
OUT 0104
END

9.7   SUMMARY

The impact of PLC technology on the manufacturing industry has been tremendous. In addition to the savings for the automotive industry mentioned at the beginning of the chapter, there is also the considerable flexibility offered to that segment of industry that manufactures piece parts. Traditionally, these industries have produced batches of parts. The time and cost of changing the process to produce a different part was so expensive and time consuming that it was not feasible to turn out small batches. With PLC control, it is often feasible to insert the requirement for a single piece into the production run.

For example one gear manufacturer uses a PLC controlled, robot-operated manufacturing cell concept. The gear blanks are fed to the cell on a conveyor. Each machining process is controlled by the PLC, as is the robot, which moves the gear blanks from machine to machine. Each blank is treated as a production batch of one, responding to the processes programmed into the PLC. Therefore, if a customer needs a single gear or a thousand of the same type, it is a simple matter of telling the PLC how many times to run the process program and then ensuring that the blanks are fed to the cell in the appropriate order.

Programmable controllers offer increased flexibility in industrial automation over common relay logic and control systems. PLCs are specially designed computers that are "hardened" for application in the often hostile environments presented by the plant floor or mobile equipment installations. They consist of a central processing unit, a memory unit, input and output modules, and an operator interface.

Programming of a PLC can be accomplished through a dedicated keyboard or a personal computer, or both, depending on the particular PLC. Programming languages vary from low-level languages such as ladder logic and Boolean mnemonics to high-level languages such as PASCAL and FORTRAN. A more recent development in PLC programming is the GRAFCET method, which uses a graphical representation of the process sequence that can then be translated into the syntax of numerous PLC types automatically through the use of compiler software such as CADEPA.

Several application examples were discussed, through which you learned to read and construct ladder logic diagrams and to write the Boolean mnemonic program for a specific PLC. You will have the opportunity to practice this new skill by completing the exercises at the end of the chapter.

REFERENCES

Baracos, Paul. 1992. GRAFCET Step by Step. St-Laurent, Quebec: Famic Automation, Inc.

Childs, James J. Jr. 1987. *Programmable Controllers—What Are They and How Are They Used?* Commline (Summer).

Stubbins, Warren Fenton. 1986. *Essential Electronics*. New York: Wiley.

REVIEW QUESTIONS

General

1. List and describe the functions of the five elements of PLC architecture.
2. Define and explain the following terms:

   RAM

   ROM

   PROM

   EPROM

3. Describe a transistor and explain the terms NPN and PNP as they relate to transistors.

Circuit Practice

Draw complete electrohydraulic control circuits (hydraulic and ladder logic diagrams) to perform the following operations and write the representative Boolean mnemonic program. If software such as Automation Studio is available, use it to draw and simulate the system. Circuits are to include momentary push buttons, limit switches, and other inputs where automatic action is required.

4. Use a single-solenoid DCV to cause a double-acting cylinder to cycle one time automatically.
5. Use a single-solenoid DCV to cause a double-acting cylinder to cycle continuously. The operation must be fully automatic.
6. Use a single-solenoid DCV to cause a double-acting cylinder to cycle automatically for 10 cycles, then stop in the retracted position.
7. Use a single-solenoid DCV to cause a double-acting cylinder to cycle automatically for 10 cycles with a 5-s delay between cycles.
8. Repeat Problem 5 using a two-position double-solenoid DCV.
9. Repeat Problem 6 using a two-position double-solenoid DCV.
10. Repeat Problem 7 using a two-position double-solenoid DCV.
11. Use a three-position double-solenoid DCV to cause a double-acting cylinder to extend, hold for 3 s, retract, hold for 5 s, then begin the cycle again. The cycling must continue automatically for 100 cycles or until a STOP button is pushed.
12. Three cylinders are required to operate sequentially. Design circuits to cause the cylinders to operate in the following sequences:

    a. X1, X2, X3, R random
    b. X1, X2, X3, R3, R2, R1
    c. X1, X3, R3, X2, R2, R1
       (X = extend, R = retract)

13. Use a proportional control valve to cause a bidirectional hydraulic motor to operate with two forward speeds and two reverse speeds. It must operate at each speed for 10 s. It must stop after completing the second reverse speed.

CHAPTER 10

Robotics

OBJECTIVES

When you have completed this chapter you will be able to:

- Explain the differences between robots and automated machines.
- Discuss the five basic components of a robot.
- Explain the five basic geometries of robot manipulators.
- Understand the term *end effector* and discuss the different types of grippers.
- List the three robotic power systems and discuss the applications of each.
- Explain the differences between servo-controlled and non-servo-controlled robots.
- Define *point-to-point*, *controlled path*, and *continuous path*.
- Explain the three methods for programming a robot controller.
- Discuss methods used to ensure personnel safety when programming, maintaining, and operating robots.

10.1 INTRODUCTION

To this point we have discussed the various elements that make up what is generally referred to as *motion control*. The discussion has included the power system (specifically, hydraulic power, although most of the discussion would apply to pneumatic power as well), the control system, and the related sensors and transducers. The control systems discussed—direct switching with limit switches and other discrete elements, PLCs, and operational amplifiers—are as applicable to pneumatic and electrical drive systems as they are to hydraulic drives. Likewise, the sensor and transducer technologies are equally applicable, regardless of the power used to operate the device.

Through example circuits we have seen many concepts of motion control or automation without ever actually introducing the subject. In this chapter we discuss motion control in its most glamorous (although not always its most sophisticated or technologically demanding form)—robotics.

There are five basic components or groups of components that make up a robot—the manipulator, the valves and amplifiers that operate the manipulator, the power source, the controller (including the computer and the feedback device), and the end effector. After attempting to define a robot (a difficult task in itself), we will look at each of these sections in some detail.

10.2   ROBOTIC DEVICES

Although some robotlike machines were developed and used in Europe in the 1920s, robots and robotics have long been associated with science fiction, in which robots were (and still are) depicted as semi-intelligent (but often mindless) machines that had at least some human characteristics and, usually, a roughly human form. The term *robot* was introduced in 1920 in the play *Rossum's Universal Robot*, written by Karel Capek. The word was derived from the Czech word *robota*, meaning "compulsory labor." In this play, Rossum (the hero) invented anthropomorphic machines to be his servants.

In the 1940s, science fiction writer Isaac Asimov introduced robots onto the American and, subsequently, international science fiction scenes with numerous books, beginning with *I, Robot*. He coined the word *robotic* and penned the *three laws of robotics*, which were recognized for many years as the defining principles for robot behavior:

1. A robot must not harm a human being or, through inaction, allow a human being to come to harm.
2. A robot must always obey a human being unless this is in conflict with the first law.
3. A robot must protect itself from harm unless this is in conflict with the first or second law.

After a few years, the more sinister forces came on the scene and devised robots after their own evil image, defying Asimov's three laws and using their robots to wreak havoc on humanity. As a result, we now see robots running the moral gamut from R2D2 and C3PO of *Star Wars* and Data of *Star Trek: The Next Generation* to the Cyborg (*Battle Star Galactica*), the Borg (*Star Trek*), and Data's evil twin brother, Lore.

Although these robots are all fictional representations, they represent—at least to some extent—a part of the future of automated machines. We are already seeing machines with motions that closely resemble human motion. Universal Studio's ET and King Kong are exquisite machines in which finger movement and muscle flexure are extremely realistic. We also see machines that have a good deal of decision-making capacity, although they are incapable of *thinking* in the normal meaning of the word. Drawing recent international attention have been chess matches between computers and world-renowned chess masters that have shown that computers can play chess "with the best of them." These trends indicate that the appearance of at least a crude version of those fictional robots is not too far off.

Although these devices are interesting, exciting, glamorous, and, to some extent, frightening, they are not of overwhelming interest to either the manufacturing or service industries or to motion control specialists. The more mundane processes of machining, welding, spray painting, and other procedures to put a product out the door in greater quantities, with higher quality, and at lower cost are more likely to be of concern today.

An *industrial robot* is often defined as a reprogrammable, multifunction manipulator designed to move material, parts, tools, or specialized devices through variable programmed motions to perform a variety of tasks. You recognize immediately that this is a somewhat vague definition. In fact, there are no indisputable, sharp boundaries in defining robots and robotics. The definition indicates that a robot must be "reprogrammable" but does not define that term; thus, you may logically wonder if it means that the device must be computer operated, or if simply moving a limit switch constitutes "reprogramming." Some "roboticists" insist that a robot must possess some degree of intelligence, necessarily implying the inclusion of a variety of sensors and a somewhat sophisticated computerized controller.

For our purposes, let's *begin* with this definition of a robot, then expand it to include, "A robot must have the capability of independently (without external intervention) performing 'mental' and physical tasks, and it must possess some degree of intelligence."

This definition narrows the field somewhat but still does not draw any firm lines of demarcation that would allow you, in all cases, to say, "This is a robot," or—more difficult, perhaps—"This is *not* a robot." To give you a little more help with this problem— but certainly not to solve it completely—some additional definitions might be useful.

The first of these defines a device that is very closely related to a robot. In fact, it is often very difficult to determine the real difference. This device is called a *pick-and-place mechanism*. This is a mechanical handling device that is equipped with a gripper or other device to hold an object. It performs a predetermined movement according to a fixed profile. Generally, no feedback or other such information is involved, although limit switches or similar devices may be used to determine that a specific position has been reached.

A second device, termed a *teleoperator*, is frequently found providing the human touch in hazardous operations (such as handling very hot, cold, or radioactive materials) and in remote underwater operations. Teleoperators or *telemanipulators* are devices that respond to a human input. They are commonly controlled by switches, push buttons, or joysticks, although technology is currently available that will enable such machines to directly replicate the movement of the operator's hands and fingers.

The humanoid robots of science fiction, by the way, are correctly termed *androids*. An android always resembles a human in appearance; however, it may far exceed the physical abilities and pure logic and "intelligence" of its human model.

The Borg of *Star Trek: The Next Generation* is yet another humanoid mechanism that is referred to as a *cyborg*. A cyborg is part machine and part biological organism. The term is a combination of the words *cybernetics* (the science of control systems) and *organism*.

Although some confusion probably remains about what a robot really is, simply adapt the remainder of this chapter to your ideas of robotics.

10.3   MANIPULATORS

The *manipulator* is the first thing we see when we look at a robot. It is the moving structural hardware of the device. It may be as simple as a single cylinder that moves a gripper, or it may contain multiple moving elements capable of moving through some very complex motion patterns. The manipulator does not include the end effector.

The *geometry* (or *kinematics*) of a manipulator defines the type of motion of which the robot is capable. There are five defined geometries—Cartesian, polar, cylindrical, anthropomorphic, and SCARA. These are illustrated in Figure 10.1.

10.3.1 Cartesian Geometry

A robot that operates with a *Cartesian geometry* is limited to straight-line motion. The Cartesian robot illustrated in Figure 10.1a moves in three axes (*x*, *y*, and *z*). One- or two-axis motion is also used. The *envelope* of motion (the volume within the reach of the robot) is a cube defined by the minimum and maximum travel of each element. The end effector can be placed at any point within that cube. All motion is rectilinear; there is no rotary motion.

Rather than using the pillar configuration shown in Figure 10.1a, the mechanism may be suspended from a *gantry*. This configuration is shown in Figure 10.2. The gantry provides the most rigid structure of all robot configurations.

The Cartesian robot allows for the most simple control algorithms of all the possible configurations. Since all motion is rectilinear, the mathematical representation of positions is uncomplicated. Also, because of the linear translation, the orientation of the end effector remains unchanged, both in space and in relation to the manipulator throughout the envelope. This further simplifies the control algorithm.

10.3.2 Polar Coordinates

Figure 10.1b illustrates a robot with *polar* (or *spherical*) coordinates. This type of unit has two rotating joints and one translating joint and is sometimes called an RRT configuration (R for rotation, T for translation). Its axes are defined as R, θ, and ϕ, as shown in the drawing. The envelope of this device is, theoretically, a spherical shell whose thickness is defined by the range of travel of the translating arm. For practical reasons, the entire theoretical envelope is seldom realized.

10.3.3 Cylindrical Coordinates

A *cylindrical* robot uses one rotating joint and two translating joints (RTT), as shown in Figure 10.1c. The axes are defined as r, z, and θ. The envelope is theoretically a cylindrical shell, although the rotation is usually limited to about 300°.

10.3.4 Anthropomorphic

The robot shown in Figure 10.1d is called *anthropomorphic*, meaning "jointed like a human." It is also termed *jointed, revolute,* or *articulated*. All three joints rotate on this type of unit (RRR), with no translation or linear motion. The joints are commonly referred to

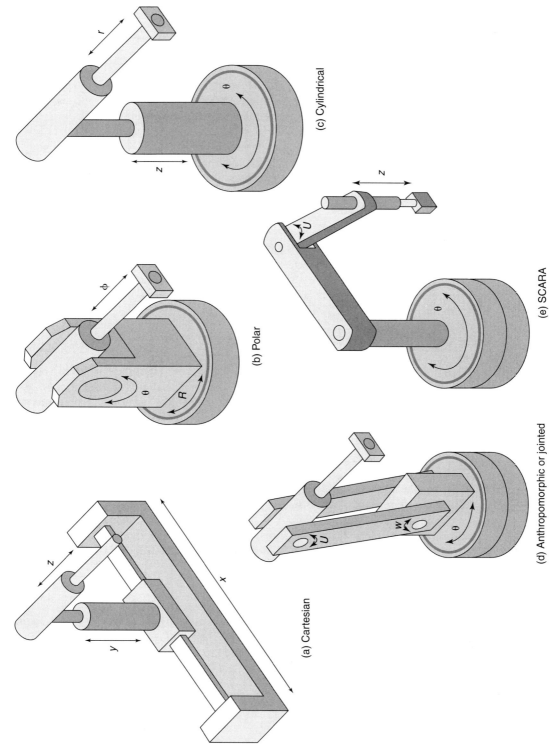

Figure 10.1 The five basic robotic manipulator geometries are (a) Cartesian, (b) polar, (c) cylindrical, (d) anthropomorphic, and (e) SCARA.

(a) Cartesian

(b) Polar

(c) Cylindrical

(d) Anthropomorphic or jointed

(e) SCARA

Figure 10.2 The gantry provides the most stable robot platform. (Courtesy of phd, Inc.)

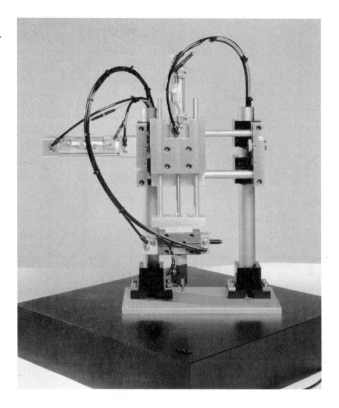

as the waist, shoulder, and elbow. There is some disparity in the designation of the coordinates, although the waist joint is usually designated θ. The shoulder joint is usually designated either ω or W, and the elbow is normally U.

The flexibility of the anthropomorphic arm makes this one of the most popular robots used for industrial applications. It is compact compared with other robot types, which means that it has a large working volume in relation to its size.

10.3.5 SCARA

The term *SCARA* is an acronym for Selective Compliance Assembly Robotic Arm. As shown in Figure 10.1e, the SCARA robot has two rotating joints and one translating joint (RRT). In these devices, the rotation of both joints is about the vertical axis. The rotational coordinates are usually designated θ and U. The vertical motion (designated Z) may be in either the main pillar or in the segment holding the end effector.

SCARA robots are frequently found in assembly operations, especially in the electronics industry. Typically, these units are very fast but can carry only very light payloads.

10.3.6 Other Configurations

Configurations other than those discussed in the preceding section are possible for special applications; however, as more axes (hence, degrees of freedom) are added, the control algorithms become much more complex. The usual approach is to use the standard robot configurations and perform any special maneuvers using the end effectors. The manipulator is actually just the vehicle for delivering the end effector to some precise spatial reference point. Once the end effector reaches that point, it is usually required to perform some task on its own. This may be as simple as extending a small linear actuator to stamp a part, or as complex as maneuvering a multifaceted part to insert it precisely into a receiver without touching the sides of the hole. This may require considerable dexterity on the part of the end effector while the manipulator sits and waits for the task to be completed. Depending on the task for which it is intended, the end effector may be the most demanding part of the robot design in terms of actual design, manufacture, and control. For most applications there exist off-the-shelf manipulators to deliver the end effector to the desired location. Thus, the real challenge is to make the end effector perform its task once it is in position.

10.4   END EFFECTORS

The *end effector* is the device that is attached to the end of the robot arm to accomplish the designated task. It is sometimes called the end of arm tool (EOAT). The types and complexity of end effectors are virtually limitless. They may be for gripping, machining, painting, inspecting, cleaning, or virtually any other task. Some typical examples are shown in Figure 10.3. The most common (and perhaps the most interesting) of these devices are the grippers.

As the name implies, a *gripper* is a device used to grasp or grip an object. Although magnetic and vacuum grippers, along with inflatable bags for delicate items, are sometimes used, by far the most common type uses mechanical fingers. The fingers are opened and closed using hydraulic, pneumatic, or electrical actuators.

The two-finger angular closing gripper shown in Figure 10.4 is the most simple and most common type of gripper. It uses a short-stroke, single-acting pneumatic cylinder to close the fingers and a spring to open them, or vice versa. In reality, the device is merely air-operated pliers.

Figure 10.3 Examples of end effectors. (Courtesy of Mack Corporation)

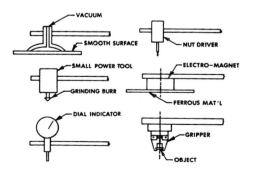

Figure 10.4 An angular gripper can be opened in several ways. This one is pneumatically opened and spring closed. (Courtesy of phd, Inc.)

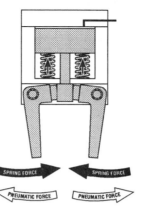

Figure 10.5 illustrates a two-finger parallel closing gripper. In this particular device, a double-acting, double-ended pneumatic cylinder is used to open and close the fingers. The parallel action means that this gripper is essentially an air-operated vise.

Two-fingered grippers work well for gripping objects with parallel sides. For other shapes, it may be necessary to devise adaptors to accommodate the lack of parallel surfaces. For example, cylindrical or spherical objects can be handled by using V-blocks that provide four contact points, as shown in Figure 10.6. Other adaptors can be used to accommodate other shapes. Some examples are shown in Figure 10.7.

Although grippers are usually used to grip the external surface of an object, it is also possible to grip the internal surface of an object. An example is a bearing that is to be inserted into an inside cavity. An internal, or inside diameter, gripper is illustrated in Figure 10.8

Figure 10.5 A parallel gripper is similar to an air-operated vise. (Courtesy of phd, Inc.)

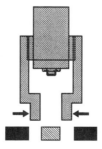

Figure 10.6 V-blocks provide a four-point contact for gripping spheres and cylinders.

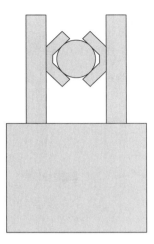

Other methods of mechanically gripping items with no parallel sides include the use of three- and four-finger mechanisms. Again, adaptors may be required for some shapes or object orientations. Two-, three-, and four-finger grippers usually have the self-centering capability of a universal chuck to ensure the correct alignment and orientation of the object every time.

Much more complex gripping devices are possible. These usually involve the incorporation of additional joints in the fingers themselves. One example is a jointed two-finger gripper in which the joints are individually actuated. This allows the same gripper to be used for a variety of shapes and sizes. Even more complex are the anthropomorphic hands, such as those used in Universal Studio's ET at their theme park in Florida.

Figure 10.7 Grippers can be modified to handle virtually any shape.

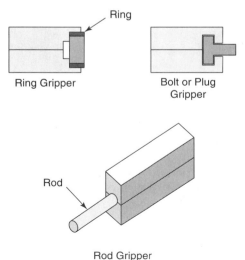

Figure 10.8 Grippers can also be designed to grip parts internally.

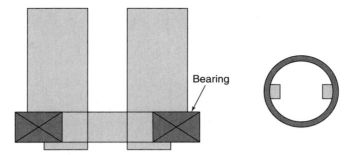

Bearing

Although the mechanical grippers just discussed can be equipped with tactile sensors to ensure that they do not crush or otherwise damage the part during handling, it is sometimes advantageous to use devices that will provide more finesse. Popular options include pneumatically inflated devices, vacuum devices, and magnetic devices.

Pneumatic devices can be used either internally or externally. Figure 10.9 shows two such devices. On the left is an inflatable tube that can be used to grip virtually anything into which it can be inserted. The inflatable collar on the right is used to surround the item and grip it evenly at multiple contact points. Neither device is limited to round objects. The inflatable portion of the gripper can be constructed to accommodate virtually any shape.

The ability of the pneumatic devices to firmly grip and hold an object depends on several factors. The most important of these are the surface finish and shape of the object, the surface finish of the material used in the gripper, and the inflation pressure. Since the tube and collar are actually very rugged balloons, they cannot be expected to provide rigid support. However, the rigidity can be increased somewhat by reducing the volume

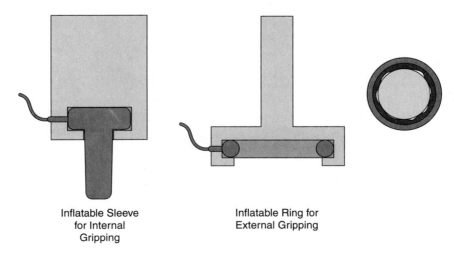

Inflatable Sleeve
for Internal
Gripping

Inflatable Ring for
External Gripping

Figure 10.9 Inflatable grippers can be used for fragile parts and can compensate for misalignment.

Figure 10.10 Schematic of a vacuum gripper with "blow-off" capability.

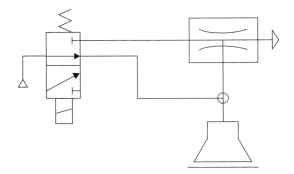

of material to be inflated. For example, a rod can be inserted in the inflatable tube to decrease the total volume of air required. This lack of rigidity can be an advantage. In cases where the part is being mated with another part, the flexibility of the gripper can accommodate minor misalignments.

A vacuum gripper is shown schematically in Figure 10.10. In this unit the directional control valve is used alternately to initiate the vacuum, then break the vacuum and direct pressurized air into the vacuum cup. This is termed a *blow-off* system.

A typical three-axis end effector is shown in Figure 10.11. This type of device is called a *wrist*. It incorporates motion about the three axes, which are termed *pitch*, *roll*, and *yaw*. These axes are designated β, α, and γ, respectively. In the design or selection of a wrist, several items must be considered. The weight of the wrist must be subtracted from the maximum payload of the manipulator; thus, weight may be a selection factor. Likewise, the volume and operating range and envelope may be critical because of the

Figure 10.11 A three-axis wrist requires a very complex control program to position it correctly.

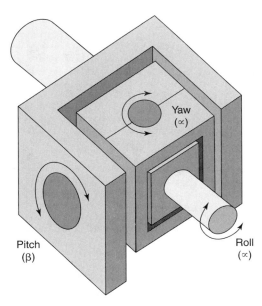

Pitch
(β)

Yaw
(α)

Roll
(α)

Figure 10.12 A robot with five axes or degrees of freedom. These do not include the gripper movement.

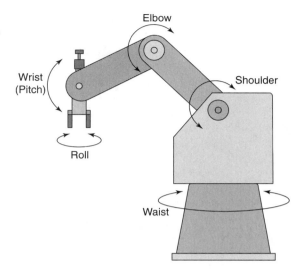

clearances in which it must operate. It must provide for the actuation of the tool it carries—gripper, drill head, electromagnet, or whatever. Finally, it may require some degree of compliance or flexure to compensate for slight misalignment when performing operations such as inserting pins. This may include force or tactile pressure sensing.

The number of axes about which a mechanism can move is expressed as its *degrees of freedom*. The robot depicted in Figure 10.12 has movement in five axes and thus has five degrees of freedom. The operation of a gripper or other tooling is sometimes considered an additional degree of freedom. If the robot shown was mounted on a slide or trolley to allow it to service more than one work station, each direction of motion would constitute another degree of freedom. The significance here is not in the complexity of the machine itself—the actual mechanism may be relatively simple—but in the *control* of the machine. We'll discuss this problem in later sections of the chapter.

10.5   POWER SYSTEMS

The power system, or prime mover, for a robot can be hydraulic, pneumatic, or electrical. Most early industrial robots used fluid power—either hydraulics or pneumatics—to power the manipulator. Subsequent developments in electrical drives combined with computer controls and, later, servomotors gave rise to the electrical robot, which supplanted many of the early hydraulic and pneumatic units.

As with many other factors concerning robots, there are no firm rules concerning which drive system is best for a particular application. Such decisions must be based on the characteristics of the particular task, including weights to be handled, operating envelope, speeds, accuracy and repeatability required, and many other parameters of the task, environment, and even personal preferences.

By current estimates, 50 to 60% of the industrial robots operating today use electrical drives. These units may use either AC or DC motors connected to the manipulator components through various types of linkages, gear reduction systems, belt and chain drives, ball screws, lead screws, and others. Electric drive motors—particularly stepper motors and servomotors—lend themselves readily to very precise computer control. They can provide very smooth acceleration and deceleration of the load. Using gear reduction systems, electric motors can deliver the high torque needed to handle heavy loads. Recent developments in brushless DC drives allow for precise speed control. Because they do not produce sparks from brushes riding on a commutator, they can also be used in some hazardous areas where electric motors previously were forbidden.

Hydraulic drive systems are used in 20 to 25% of today's industrial robots. They are used in applications where large forces are required and where heavy objects must be moved and positioned with accuracy and precision. Generally, the manipulator components are driven directly using hydraulic cylinders or low-speed, high-torque hydraulic motors. This direct-drive capability eliminates the various drive mechanisms required by many electrical drives and provides for precise handling of heavy loads. Hydraulic drives have the highest torque capabilities of the three drive options.

Pneumatic drives are limited to small robots because of their very limited power capabilities. Their big advantage lies in the very high speeds available from small pneumatic cylinders. Although the compressibility of the air is often considered a liability, it can be used to advantage in cushioning the impact at the end of the stroke and in providing a degree of finesse, particularly in the gripper action. Pneumatically driven air motors have very limited torque capabilities, even when used with gearboxes.

Pneumatic drives are the least expensive of the three alternatives. They are easily controlled by computer systems, although their accuracy and precision is degraded somewhat by the compressibility problem in some applications. The compressed air used in robots generally must be far cleaner, drier, and better regulated than that found in most shop applications. Well-prepared air will ensure the most precise operation and longest life from the pneumatic components.

10.6   CONTROL

As we discussed earlier, the control of the robot may present a greater challenge than the mechanism. Any increase in the complexity of the machine is likely to result in a much more significant increase in the complexity of the control system, particularly in the computer algorithms. Figure 10.13 shows the architecture of a robot control system. Here we see that there are five basic sections—the manipulator or structure (including the actuators that move the manipulator members), the valves and amplifiers that operate the actuators, the power supply, the control system (which includes the controller and the feedback devices), and the end effector. The type and sophistication of valves, feedback devices, and the controller depend on the operating mode and the degrees of freedom of the robot.

Robots can be classified as *servo-controlled* and *non-servo-controlled*. The non-servo-controlled units are the least complex from the control point of view. They are also called *point-to-point* robots and usually fall into the category of *pick-and-place* mecha-

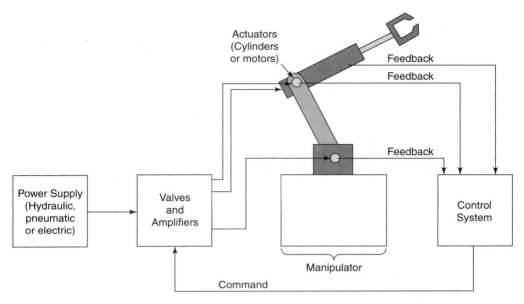

Figure 10.13   Basic robot architecture includes the manipulator, power supply, control system, and end effector.

nisms. In the operation of these devices, only the end points of their motion are of any interest. How they get to those end points is of little, if any, concern (as long as they do not bang into something). The end points are sometimes defined by hard mechanical stops, but usually there is some type of discrete position sensor such as a mechanical limit switch, electro-optical sensor, or some similar detector. Some purists feel that these units are not really robots, because they lack the "reprogrammable" element found in the basic definition of a robot. In most cases, however, they are considered robots.

In point-to-point robots there is usually motion in only one axis at a time. As one axis reaches its designated end point the sensing device generates a signal that ultimately causes its motion to stop and starts the motion of the next axis.

The significant difference in the control concepts of non-servo-controlled and the servo-controlled robots is the interest of the controller in the whereabouts of the end effector. In the nonservo units, the controller does not know (and does not care) where the end effector is until the designated points are reached. In the servo units, the location of the end effector is always known to the controller; thus, continuous position sensors are required.

Servo control systems fall into two categories—controlled path and continuous path. A *controlled path* system is, in reality, a point-to-point system, because the end points are of more concern than the actual path between the points. However, the controller knows the location of the end effector at all times; therefore, the motion of all axes can be co-ordinated to allow for a straight line between points. Any number of points can be programmed to allow the robot to work around obstacles; however, the path between any two designated points will always be a straight line.

A *continuous path* control system allows the end effector to follow a smooth path of any configuration. An example of such a device is the robot that applies the sealer around an automobile windshield before it is placed on the vehicle. Welding and paint-spraying robots also use this system. In reality, continuous path is nothing more than a very extensive controlled path system with a very large number of points. As you can imagine, this requires a great deal of memory in the controller as well as a very accurate position feedback system.

Both controlled path and continuous path control systems require continuous position feedback from all axes. For rotary components, angular position can be measured simply and inexpensively using potentiometers that provide a continuous analog signal; however, these devices are less accurate than the digital signals provided by either the photoelectric encoders or the electromagnetic resolver. Linear position feedback can be provided by linear potentiometers, LVDTs, linear encoders, or a variety of magnetic or electrical resistance sensors. Both rotary and linear position transducers are discussed in detail in Chapter 8.

10.7   PROGRAMMING

For a robot to perform its prescribed task, its controller must be programmed to tell it what to do and when to do it. There are three methods of programming: off-line programming, pendant teaching, and lead through. In off-line programming, the instructions for the robot are physically typed on a keyboard, usually using a specialized, robot-specific programming language. These languages are usually similar to PASCAL but contain special terminology and syntax. Separate lines in the program will contain instructions for the target position, dwell or delay time, speeds, and other information required to complete the task. In the case of non-servo-controlled, point-to-point robots, the program may be a few lines of instruction for a small PLC. For a continuous path robot, thousands of lines of input may be required.

The most common method for programming point-to-point servo robots is through the use of a *teaching pendant* such as the one shown in Figure 10.14. The pendant is a handheld box with buttons, switches, or joysticks used to control each axis of the robot. Using the pendant, the operator moves the robot to a desired point, then presses a button that transfers the information from each position sensor to the controller memory. The operator then moves the robot to the next position and enters it. In this way, the operator "teaches" the controller the desired points.

A problem can arise in this process in that the controller does not know the path the operator used to reach any specific point. As a result, it simply takes the shortest path from point to point. If it is important that a point be approached in a specific way, a series of approach points may have to be used. The actual number of points used in a procedure is limited only by the memory size of the controller. Many teaching pendants allow the operator to put in additional control information such as speed, dwell time, gripper opening and closing, and other process data.

For continuous path operations, the controller is "taught" the path by "observing" the operation of the manipulator as it is manually moved through the operation. For example, a human welder may physically grasp the end effector on a welding robot and run a bead along a part. With the controller in the teach mode, the data from the position sensors are continuously recorded. This recorded information becomes the control program. It likely will

Figure 10.14 A teaching pendant can be used to program a servo-operated robot.

include, in addition to path data, instructions on speed, dwell time, when the welder is on and off, and so forth. In cases where the actual manipulator is too heavy or dangerous to be "led by the nose," a lightweight and unpowered but otherwise exact replica can be used.

10.8   SAFETY

Despite Asimov's original rules of robot behavior, today's industrial robots can unintentionally cause injury and death to the careless and unwary. As a result, common sense and government regulations require that safety procedures and devices be used to protect their human masters from accidental harm. Several methods are used.

The persons most likely to be injured by a robot are the maintainers and programmers, who often must be in physical contact with the robot while it is powered in order to do their jobs. In most cases, there is a switch or button on the control console that disables all other control elements on that console and allows the robot to be operated only from the teaching pendant or a similar maintenance device. The person in contact with the robot is usually required to have physical possession of the pendant so that it cannot accidentally be operated by another person. In addition, many pendants have an emergency stop button, a dead man handle, or both. These devices will completely and immediately disable and stop the robot when actuated.

When the robot is operating, devices must be provided to either prevent human encroachment on the work envelope or to shut down the robot if such encroachment occurs. Natural barriers of peripheral equipment such as conveyors, spray booths, and other devices employed in the process may be effective. Physical barriers such as wire mesh or Plexiglas cages may be required. If cages are used, safety interlocks are required to ensure that all doors are closed and locked and that all removable sections are in place before the robot can be operated.

In lieu of barriers, pressure-sensitive floor mats are sometimes used. These devices may contain a series of small air passages or fiber-optic cables. In the air units, external

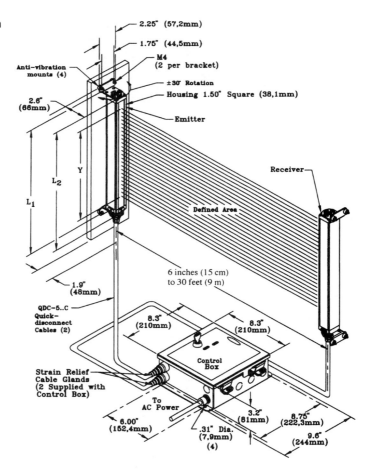

Figure 10.15 A light screen such as this one can extend for up to 30 ft. (Courtesy of Banner Engineering Corp.)

pressure (such as a foot) on the mat will restrict the airflow and cause a back pressure that is sensed by the controller. The same external pressure would cause a detectable change in light intensity in the fiber-optic devices. Both types are very fast and acceptably sensitive. However, they are fairly susceptible to damage resulting from dropped objects as well as debris from the production process.

Perhaps the most popular safety barrier is the infrared light curtain. This unit consists of an infrared emitter and receiver connected to a microprocessor, as shown in Figure 10.15. The microprocessor, in turn, interfaces with the robot controller and provides a GO/NO GO safety signal. When the light curtain is activated, any interruption of any beam will cause the robot to stop and remain inactive until a reset button is actuated.

Light curtains may be installed either vertically or horizontally. The height of the curtain can be from a few inches to several feet, depending on the dimensions of the emitter. The sensing range is from a few inches to 45 ft, depending on the particular unit. Other features include the ability to completely enclose an area through the use of mirrors, and the selective blanking of certain areas to allow permanent structures such as conveyors to penetrate the curtain. Figure 10.16 shows a robot work cell

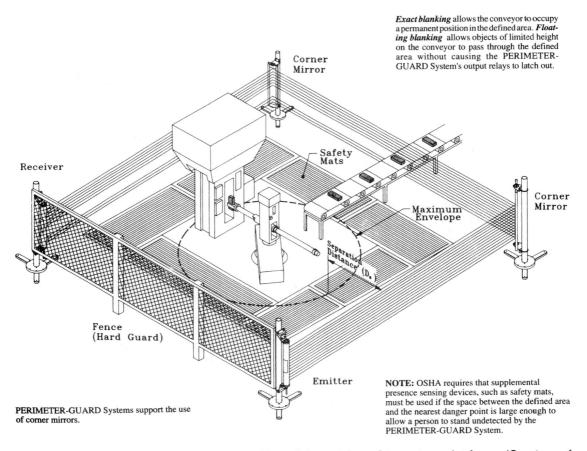

Exact blanking allows the conveyor to occupy a permanent position in the defined area. **Floating blanking** allows objects of limited height on the conveyor to pass through the defined area without causing the PERIMETER-GUARD System's output relays to latch out.

PERIMETER-GUARD Systems support the use of corner mirrors.

NOTE: OSHA requires that supplemental presence sensing devices, such as safety mats, must be used if the space between the defined area and the nearest danger point is large enough to allow a person to stand undetected by the PERIMETER-GUARD System.

Figure 10.16 A robotic workstation protected by a light curtain, safety mats, and a fence. (Courtesy of Banner Engineering Corp.)

protected by a three-sided light curtain with blanking for the conveyor, a barrier fence, and safety mats.

10.9   INSTALLATION CONSIDERATIONS

Three factors must be considered when planning the introduction of robots into the workplace. The first of these is purely technical: Can a robot do the job? Basically, any job a human can do, a robot can do. In fact, the best way to set up a robotic work cell is to have a human model go through the process.

Technical ability does not always make economic sense, however. In many cases, simple automation is a far better economic solution than robotics. Stamping out metal disks from a continuous strip, for example, is no job for a robot. Simple automation is probably more practical. Factors that favor the introduction of a robot include a shortage of skilled labor and the need to increase the quality and/or quantity of the product.

Finally, there is the social factor. Although there will always be some impact on both factory floor workers and management, the negative results can be minimized in most cases. Usually, the workers displaced by robots can be retrained and put into more productive positions. The introduction of robots has been welcomed in many situations because robots took over many dangerous and hazardous jobs as well as those jobs involving continuous, repetitive, low-level tasks that nobody liked to do. In some circumstances the introduction of robots actually saved jobs by allowing a company to remain competitive (or even in business) by reducing costs, increasing production, and improving quality. In all situations the feelings of and economic impact on the workers must be considered if the introduction is to be totally successful and well received.

10.10   SUMMARY

The accepted definition of a robot requires that it must be reprogrammable. That definition immediately eliminates numerous automated machines but leaves many others in a gray zone. The primary components of a robot are the manipulator, the valves and amplifiers, the end effector, the power system, and the control system, which includes the controller and the feedback devices. Manipulators almost always fall into one of five geometries—Cartesian, polar, cylindrical, anthropomorphic, or SCARA.

The manipulator is used to deliver the end effector to a desired point in space. The end effector is basically the device that does the work. It may be a tool such as a drill or punch, or it may be some sort of gripping device. The manipulator, as well as the end effector, can be powered electrically, hydraulically, or pneumatically.

Both the manipulator and (usually) the end effector are controlled by a computer. The robot's path may be defined as non-servo-controlled point-to-point, servo-controlled continuous path, or servo-controlled point-to-point (controlled path). Regardless of the path, the controller must be programmed to allow it to direct the manipulator to the desired points. Programming may be accomplished by typing in the data in a prescribed programming language, by using a teaching pendant, or by physically moving the device from point to point. Servo-controlled robots require continuous feedback so that the controller knows the exact position of the end effector at all times.

Safety must be a high-priority concern in teaching, maintaining, and operating robots. Detailed safety procedures are required when programming and maintaining robotic systems, and physical barriers, safety mats, and light curtains are used during normal operation.

SUGGESTED ADDITIONAL READING

Hodges, Bernard. 1992. *Industrial Robotics*, 2d ed. Oxford: Butterworth-Heinemann.
Malcolm, Douglas R. Jr. 1988. *Robotics, An Introduction*, 2d ed. Boston, Mass.: PWS-KENT.
Todd, D. J. 1986. *Fundamentals of Robot Technology*. New York: Halsted Press, a division of Wiley.
Warnecke, H. J., and R. D. Schraft. 1982. *Industrial Robots Application Experience*. Bedford, England: IFS.

REVIEW QUESTIONS

1. Define the term *robot.*
2. List and define the five basic components of a robot.
3. List and define the five basic geometries of robot manipulators.
4. List the three methods of powering manipulators, and discuss the applications of each.
5. Explain the differences between servo-controlled and non-servo-controlled robots.
6. Define point-to-point, continuous path, and controlled path.
7. What are the three methods of programming a robot?

APPENDIX A

Common Fluid Power Symbols

Included in this appendix are commonly used symbols for fluid power components. This information is reproduced by permission of National Technology Transfer, P.O. Box 4558, Englewood, CO 80155.

MISCELLANEOUS HYDRAULIC SYMBOLS

Pressure Gauge		Separator with Automatic Drain	
Temperature Gauge		Filter - Separator with Manual Drain	
Flow Meter		Filter - Separator with Automatic Drain	
Electric Motor		Dessicator (Chemical Dryer)	
Heat Engine (E.G. Internal Combustion)		Heater	
Hydraulic Nozzle		Cooler	
Accumulator, Spring Loaded		Temperature Controller	
Accumulator, Gas Charged		Intensifier	
Accumulator Weight Loaded		Pressure Switch	
Filter or Strainer		Lubricator Without a Drain	
Separator with Manual Drain		Lubricator With a a Drain	

MOTORS & CYLINDERS

Fixed Displacement, Unidirectional, Rotary Hydraulic Motor	
Fixed Displacement, Bidirectional, Rotary Hydraulic Motor	
Variable Displacement, Unidirectional, Motor	
Variable Displacement, Bidirectional, Rotary Hydraulic Motor	
Oscillating Hydraulic Motor	
Single Acting Cylinder	
Double Acting Cylinder	
Double End Cylinder	

PNEUMATIC SYMBOLS

Unidirectional Pneumatic Motor	
Bidirectional Pneumatic Motor	
Pneumatic Compressor Fixed Displacement	
Flow Direction, Pneumatic	
Pneumatic Nozzle	
Muffler	
Venturi	

HYDRAULIC LINES

Working Hydraulic Line		Flexible Line		Flow Direction, Hydraulic		Plug or Plugged Connection	
Pilot Line		Lines Joining		Line to Tank (Above Fluid Level)	
Drain Line		Lines Passing		Line to Tank (Below Fluid Level)		Fixed Restriction	

METHODS OF OPERATION		BASIC VALVES	
Pressure Compensator		Check Valve	
Detent		Check Valve Pilot Operated To Open	
Manual		Check Valve Pilot Operated To Close	
Mechanical		Manual Shut Off Valve	
Pedal or Treadle		Single Flow Path Valve, Normally Closed	
Bush Button		Single Flow Path Valve, Normally Open	
Lever		Pressure Relief Valve	
Pilot Pressure		Two Position Multiple Flow Path Valve (arrows show flow direction)	
Solenoid		Open Center, Three Position Directional Control Valve	
Spring		Closed Center, Three Position Directional Control Valve	
Solenoid Controlled, Pilot Pressure Operated		Tandem Center, Three Position Directional Control Valve	
Servo		Float Center, Three Position Directional Control Valve	

BASIC VALVES			
Unloading Valve Internally Drained, Remotely Operated		Deceleration Valve Normally Open	
Needle Valve		Sequence Valve Directly Operated, Externally Drained	
Pressure Reducing Valve			
Pressure Reducing And Relieving Valve			
Counter Balance Valve With Check Valve			
Temperature And Pressure Compensated Flow Control Valve With Check Valve			

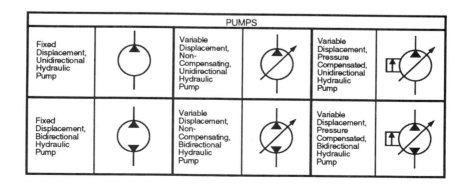

PUMPS					
Fixed Displacement, Unidirectional Hydraulic Pump		Variable Displacement, Non-Compensating, Unidirectional Hydraulic Pump		Variable Displacement, Pressure Compensated, Unidirectional Hydraulic Pump	
Fixed Displacement, Bidirectional Hydraulic Pump		Variable Displacement, Non-Compensating, Bidirectional Hydraulic Pump		Variable Displacement, Pressure Compensated, Bidirectional Hydraulic Pump	

APPENDIX B

Common Electrical Symbols

Included in this appendix are the most commonly used symbols for electrical components. This information is reproduced by permission of Womack Educational Publications.

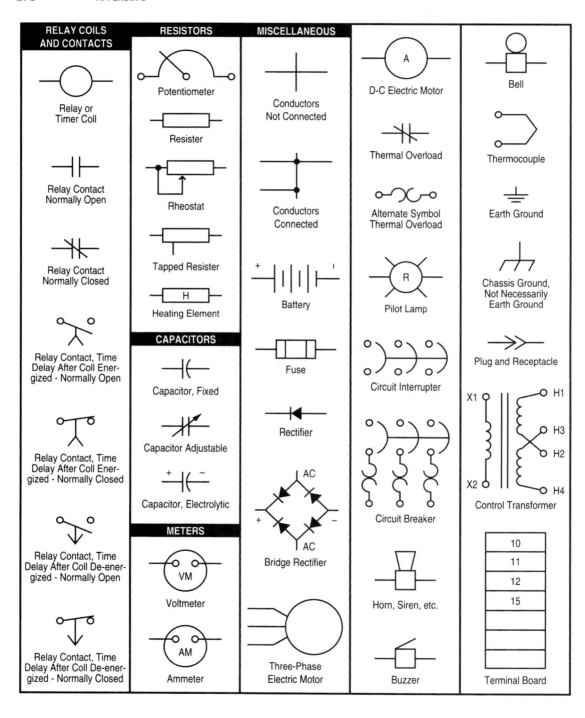

RELAY COILS AND CONTACTS	RESISTORS	MISCELLANEOUS		

RELAY COILS AND CONTACTS

Relay or Timer Coil

Relay Contact Normally Open

Relay Contact Normally Closed

Relay Contact, Time Delay After Coil Energized - Normally Open

Relay Contact, Time Delay After Coil Energized - Normally Closed

Relay Contact, Time Delay After Coil De-energized - Normally Open

Relay Contact, Time Delay After Coil De-energized - Normally Closed

RESISTORS

Potentiometer

Resister

Rheostat

Tapped Resister

Heating Element

CAPACITORS

Capacitor, Fixed

Capacitor Adjustable

Capacitor, Electrolytic

METERS

Voltmeter

Ammeter

MISCELLANEOUS

Conductors Not Connected

Conductors Connected

Battery

Fuse

Rectifier

Bridge Rectifier

Three-Phase Electric Motor

D-C Electric Motor

Thermal Overload

Alternate Symbol Thermal Overload

Pilot Lamp

Circuit Interrupter

Circuit Breaker

Horn, Siren, etc.

Buzzer

Bell

Thermocouple

Earth Ground

Chassis Ground, Not Necessarily Earth Ground

Plug and Receptacle

Control Transformer

Terminal Board

| 10 |
| 11 |
| 12 |
| 15 |

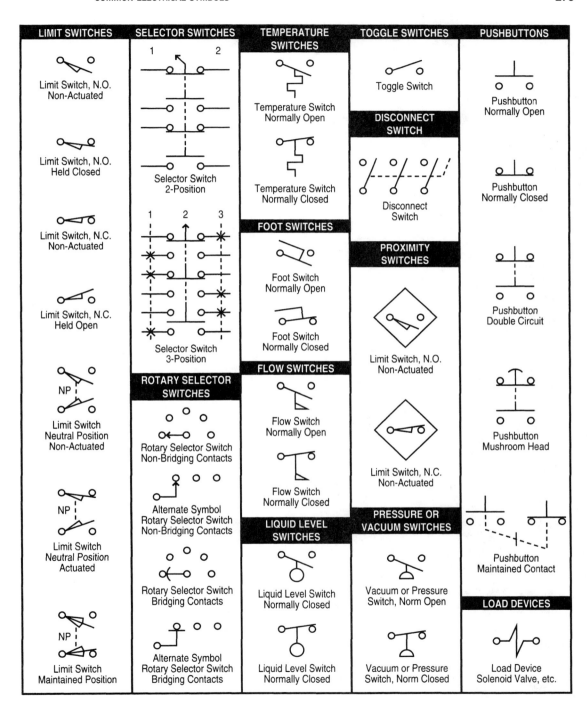

LIMIT SWITCHES	SELECTOR SWITCHES	TEMPERATURE SWITCHES	TOGGLE SWITCHES	PUSHBUTTONS
Limit Switch, N.O. Non-Actuated	Selector Switch 2-Position	Temperature Switch Normally Open	Toggle Switch	Pushbutton Normally Open
Limit Switch, N.O. Held Closed		Temperature Switch Normally Closed	DISCONNECT SWITCH	Pushbutton Normally Closed
Limit Switch, N.C. Non-Actuated	Selector Switch 3-Position	FOOT SWITCHES	Disconnect Switch	Pushbutton Double Circuit
Limit Switch, N.C. Held Open	ROTARY SELECTOR SWITCHES	Foot Switch Normally Open	PROXIMITY SWITCHES	
Limit Switch Neutral Position Non-Actuated	Rotary Selector Switch Non-Bridging Contacts	Foot Switch Normally Closed	Limit Switch, N.O. Non-Actuated	Pushbutton Mushroom Head
Limit Switch Neutral Position Actuated	Alternate Symbol Rotary Selector Switch Non-Bridging Contacts	FLOW SWITCHES	Limit Switch, N.C. Non-Actuated	Pushbutton Maintained Contact
	Rotary Selector Switch Bridging Contacts	Flow Switch Normally Open	PRESSURE OR VACUUM SWITCHES	
Limit Switch Maintained Position	Alternate Symbol Rotary Selector Switch Bridging Contacts	Flow Switch Normally Closed	Vacuum or Pressure Switch, Norm Open	LOAD DEVICES
		LIQUID LEVEL SWITCHES	Vacuum or Pressure Switch, Norm Closed	Load Device Solenoid Valve, etc.
		Liquid Level Switch Normally Closed		
		Liquid Level Switch Normally Closed		

INDEX

Answers to Selected Problems

Chapter 2

2-1. 6280 lb
2-3. F_{ext} = 21210 lb F_{ret} = 11790 lb
2-5. 7.35 gpm
2-7. 13.6 kW
2-9. F_{ext} = 78.54 kN F_{ret} = 40.06 kN
2-11. v_{ext} = 0.032 cm/s v_{ret} = 0.062 cm/s
2-13. 2.98 in^3/rev
2-15. 5.1 in^3/rev
2-17. 1151 cm^3/rev
2-19. 2.73 hp
2-21. 11.4 hp
2-23. 5 kW

Chapter 3

3-15. 5000 V
3-17. 600 W
3-19. 38.4 W   15Ω
3-21. 750 W
3-23. 0.87 W   36 mA

Chapter 4

4-11. 6.28 N   1.42 lb
4-13. 3.5 lb
4-15. 32.1 ms

Chapter 6

6-13. 7.5 V   300 mA   2.25 W
6-15. 1.74 × 10^4
6-17. 13.3 hp   564.3 Btu/min   5.5°F

Chapter 7

7-1. - 2.5 dB
7-3. 10.27 in   8.24 Hz
7-5. 9.49 in   9.78 Hz
7-7. 13.2 Hz
7-9. $\dfrac{G_1 G_2}{1 + G_1 G_2 H}$
7-11. $\dfrac{G_1 G_2 G_3}{1 + G_2 G_3 H_1 H_2 + G_1 G_2 H_3}$
7-13. 0.25 V/deg
7-15. 60 rpm/gpm
7-17. 0.02 V/rpm
7-19. 11.16 gpm
7-21. b. 0.499 in/V
       c. 1.23 in
       d. 1.5 in